FISCAL RETRENCHMENT AND URBAN POLICY

Volume 17, URBAN AFFAIRS ANNUAL REVIEWS

INTERNATIONAL EDITORIAL ADVISORY BOARD

ROBERT R. ALFORD
University of California, Santa Cruz

HOWARD S. BECKER
Northwestern University

BRIAN J. L. BERRY
Harvard University

ASA BRIGGS
Worcester College, Oxford University

JOHN W. DYCKMAN
University of Southern California

T. J. DENIS FAIR
University of Witwatersrand

SPERIDIAO FAISSOL
Brazilian Institute of Geography

JEAN GOTTMANN
Oxford University

SCOTT GREER
University of Wisconsin, Milwaukee

BERTRAM M. GROSS
Hunter College, City University of New York

PETER HALL
University of Reading, England

ROBERT J. HAVIGHURST
University of Chicago

EHCHI ISOMURA
Tokyo University

ELISABETH LICHTENBERGER
University of Vienna

M. I. LOGAN
Monash University

WILLIAM C. LORING
Center for Disease Control, Atlanta

AKIN L. MABOGUNJE
University of Ibadan

MARTIN MEYERSON
University of Pennsylvania

EDUARDO NEIRA-ALVA
CEPAL, Mexico City

ELINOR OSTROM
Indiana University

HARVEY S. PERLOFF
University of California, Los Angeles

P.J.O. SELF
London School of Economics and Political Science

WILBUR R. THOMPSON
*Wayne State University and
Northwestern University*

FISCAL RETRENCHMENT AND URBAN POLICY

Edited by
JOHN P. BLAIR
and
DAVID NACHMIAS

Volume 17, URBAN AFFAIRS ANNUAL REVIEWS

SAGE PUBLICATIONS / BEVERLY HILLS / LONDON

Copyright © 1979 by Sage Publications, Inc.

All rights reserved. No part of this book may be reproduced or utilized in any form or by any means, electronic or mechanical, including photocopying, recording, or by any information storage and retrieval system, without permission in writing from the publisher.

For information address:

SAGE PUBLICATIONS, INC.
275 South Beverly Drive
Beverly Hills, California 90212

SAGE PUBLICATIONS LTD
28 Banner Street
London EC1Y 8QE

Printed in the United States of America

Library of Congress Cataloging in Publication Data

Main entry under title:

Fiscal retrenchment and urban policy.

    (Urban affairs annual reviews ; v. 17)
    1. Urban policy–United States–Addresses, essays, lectures. I. Blair, John P., 1947-   II. Nachmias, David. III. Series.
HT108.U7 vol. 17 [HT123]   301.36'08s [301.36'0973]
ISBN 0-8039-1242-0   79-13695
ISBN 0-8039-1243-9 pbk.

FIRST PRINTING

CONTENTS

Preface □	7
PART I: Urban Policy in an Era of Retrenchment	9
1 □ Urban Policy in the Lean Society □ *John P. Blair and David Nachmias*	11
2 □ The State of Cities in the Second Republic □ *Theodore J. Lowi*	43
3 □ The Possibilities for Urban Revitalization: Constraints and Opportunities in an Era of Retrenchment □ *Louis H. Masotti*	55
4 □ Bureaucratization of the Emerging City □ *Scott Greer*	69

PART II: The Tax Revolt and Policy Priorities — 83

    5 □ Citizen Preferences and Urban Policy Types □
    Wayne Lee Hoffman and Terry Nichols Clark — 85

    6 □ The Season of Tax Revolt □
    John L. Mikesell — 107

    7 □ Public Opinion, the Taxpayers' Revolt, and Local Governments □
    F. Ted Hebert and Richard D. Bingham — 131

    8 □ A Mayor's Perspective on the Tax Revolt □
    Henry Maier — 149

PART III: Urban Policy Areas: Prospects for the 1980s — 157

    9 □ Employment Policy in Postaffluent America □
    Gary Gappert — 159

    10 □ The Diminishing Urban Promise: Economic Retrenchment, Social Policy, and Race □
    Harold M. Rose — 183

    11 □ Anglo Retrenchment and the Hispanic Renaissance: A View from the Southwest □
    Warner Bloomberg, Jr. — 211

    12 □ Urban Education: Complex Problems—No Simple Solutions □
    Chava Nachmias — 245

    13 □ Health Care Policy: Disillusion and Confusion □
    Ann Lennarson Greer — 267

    14 □ City Personnel: Issues for the 1980s □
    David H. Rosenbloom — 285

    The Authors □ — 307

PREFACE

☐ ALTHOUGH URBAN POLICY in America has never been consistent and unambiguous, the steady growth of the public sector has directly and indirectly affected the cities. The nationalization of problems such as economic growth, energy, transportation, equal education and occupational opportunities, housing and unemployment coupled with the federal government's symbolic and fiscal commitments to tackle these issues have exerted a significant impact on cities.

At the same time, cities, and especially the older Northeastern and Midwestern ones, have entered the vicious circle of "increasing demands-decreasing revenues." These cities must increasingly offer more and better services at increasingly higher prices to their dependent populations. However, in these same cities revenues have been declining due, among other things, to the movement to the suburbs of middle-class residents and manufacturing and service industries. The gap between "increasing demands-decreasing revenues" was, to a certain extent, filled by state and federal assistance. In fact, much federal assistance bypassed the states and was given directly to the cities. Notwithstanding the positive impact of such policies on the cities, it has created a greater dependence. Cities are now more than ever dependent on other levels of government to fulfill their service and management functions.

The expanding role of government in meeting societal needs has not gone unchallenged. Traditionally the controversy centered on the "freedom versus equality" or "free-market versus planned economy" arguments. However, since the New Deal this controversy has been more ideological; the public sector has, in fact, expanded and the roles of government have increased. This process, however, can no longer be assumed to continue steadily. As the economy has entered the stagflation era; as the middle class has lowered its expectations for rapid economic and social well-being; as the optimism of the 1960s epitomized by the "Great Society" vision has given way to cynicism and lower trust in government and government action, a new reality seems to be emerging. Perhaps the term "lean society" reflects best the present trends.

This volume brings together original contributions focusing on the conceptualization, the characterization, and the possible impacts of the "lean society" on our cities. The first part of the volume contains theoretical and normative articles. The tax revolt, perhaps the most visible manifestation of the lean society, is explored in Part II. The third part centers on concrete policy areas including employment, education, race and ethnicity, health, and public personnel management.

Many of the contributions were developed for a conference entitled Urban Institutions in an Era of Retrenchment sponsored by the Department of Urban Affairs at the University of Wisconsin-Milwaukee in November 1978. We would like to thank the Graduate School of the University of Wisconsin, Mayor Henry Maier, and the conference participants for their interest and involvement. Richard D. Bingham, Chair of the Urban Affairs Department, was helpful in so many ways in bringing the conference and this volume together, that we would like to dedicate it to him.

—J.P.B.
D.N.

Part I

Urban Policy in an
Era of Retrenchment

1

Urban Policy in the Lean Society

JOHN P. BLAIR
DAVID NACHMIAS

□ IT IS INCREASINGLY APPARENT to observers of American society that urban affairs cannot be divorced from the broader context of the American polity. Urban problems, or the "urban crisis" afflicting many of the nation's central cities is America's crisis. Urban affairs affect and are directly affected by policy decisions made at state and federal level. The mutual dependence between urban America and the nation was perhaps best articulated in August 1966 by Senator Abraham Ribicoff during a Senate inquiry into the state of urban government:

> The crisis of our cities is the crisis of modern United States. Seventy percent of all Americans now live in or close to cities. The number grows each year. So the fate of the city and the future of our country are one and the same thing. [James and Hoppe, 1969: 3]

At the time, Ribicoff's words reflected the public's and the government's deep concern over its cities. The rhetoric of crisis gained a prominent place on the American public agenda, and governments at all levels initiated a great number of new policies and programs to assist the cities. The federal government set the pace and the content of urban policy in the 1960s. Through new legislation, enacted in a period of general prosperity, Congress extended both the scope and the level of direct federal involvement in the cities. Major national programs were implemented in new areas such as manpower training, and assistance was provided for established local government functions, such as mass transportation and water and sewer systems. Federal urban aid increased from approximately $3.9 billion in the early 1960s to over $26 billion in the early 1970s. Furthermore, in many cases the federal government bypassed other levels of government in developing and implementing urban programs, thus enhancing the mutual dependence between cities and the national government. The implementation of programs such as Community Action and Model Cities took the form of establishing channels of funding that go directly from the national government to local communities. And with policies such as the Economic Opportunity Act of 1964, Congress broke with tradition by authorizing aid to localities for a relatively unrestricted range of functions.

As a result of these policies cities are now, more than ever, dependent on political institutions other than local governments to fulfill their traditional functions as well as those added to them in the 1960s. Ironically, however, this greater dependence comes at a time when the nation is experiencing severe economic problems; the public is distrustful and most critical of its national political institutions and leadership, and the wisdom of government intervention in the affairs of the city is being challenged on normative and empirical grounds. Future urban policy will be formulated within the context of this new reality. The present chapter examines some critical aspects of this transition and their plausible impacts on cities.

GOVERNMENT EXPANSION AND URBAN AFFAIRS

The urban policies of the 1960s have accelerated the rate of mutual dependence between cities and other levels of government. This, however, is not a simple derivative of the unique conditions that faced the cities and the nation at that time, but the consequence of a more fundamental transition in the role of the public sector in American society. Government intervention in the market economy has been growing for the last century, as have its size and scope of activities.

This, in turn, has led to a greater concentration of power at the national level as well as a greater dependence of the entire polity (including cities) upon the public sector. In fact, based on the rates of government expansion in the last century, it has been projected that by the year 2000 government will take about 45% of the Gross National Product in taxes and employ 25% of the labor force. If the exponential growth continues, by the year 2100 "the government will take everything, all of GNP, in taxes. Although the government then will employ about 50% of the labor force, everyone in effect will 'work' for the government" (Meltzer and Richard, 1978: 114). Meltzer and Richard's data do not include the significant, although less quantifiable increases in regulation.

Although there is no single explanation that best accounts for government growth, three general theories have certain explanatory power not only for the growth in size and scope of most governmental levels, but also for the growth of city government.

The idea that in democracies the public sector expands at a higher rate than the private sector was systematically introduced by Adolph Wagner some ninety years ago. Wagner expounded the law of the "expanding scale of state activity," that is, government activity increases faster than economic output. Wagner attributed this proposition to several factors: increasing regulatory services required to coordinate a more differentiated complex economy; increasing involvement of government in economic enterprise; increasing demands and pressures for public goods and social services, such as education, public health, protection, and welfare. Focusing specifically on all government expenditures in the United States, it has been observed that they grew from $1.7 billion in 1902 to over $350 billion in the early 1970s. Significantly, government spending has grown much faster than the population and inflation.

Government expansion, however, has not accelerated steadily as in other Western democracies, but in spurts during crisis periods. Wars and other severe crises have led to significant increases in government activity. National emergencies have provided the opportunities for government to expand the scope and the magnitude of its activities. The termination of a crisis has contracted government expansion, but not to its preemergency levels. Postemergency expansion has leveled out on a higher plateau than preemergency activities. As Mosher and Poland (1964: 29) summarized the American experience "expenditures of American governments has increased substantially by every measure, but these increases have occurred spasmodically with the occurrences of international and domestic crises." The "crisis-expansion" thesis appears to have strength as an explanation for the numerous social programs initiated by federal and local governments in response to the civil disorder and violence of the 1960s. Indeed, the National Advisory Commission on Civil Disorder, appointed in July 1967 by President Lyndon B. Johnson to investigate the riots and their causes, recommended massive federal aid programs in employment, education, welfare, and housing. In the Commission's words;

> These programs will require unprecedented levels of funding and performance, but they neither probe deeper nor demand more than the problems which called for them. There can be no higher priority for national action and no higher claim on the nation's conscience.

The "crisis-expansion" thesis has been criticized on the grounds that it is insufficient for explaining the public's and the authorities motivation to expand government. Nor does this theory explain the rates of growth over the last century. A complimentary thesis advances the argument that in democracies the public sector expands because of the significant disparity between the political process and market mechanisms:

> The market produces a distribution of income that is less equal than the distribution of votes. Consequently, those with the lowest income use the political process to increase their income. [Meltzer and Richard, 1978: 117]

Candidates for elective offices and elected officials attempt to attract voters with incomes near the median income distribution by offering redistributive policies and programs that impose a net cost on those individuals with higher incomes. As the franchise is extended downward through the income distribution to include more and more groups, and as lower income groups become aware of their political resources, the proportion of votes going to candidates promising redistributive policies increases. Conversely, the public sector is contracted when the net benefit from redistributing declines.

The "voter-attraction-redistribution" thesis has implications for government allocative and regulatory activity, as well as for pure transfer policies because income distribution may be affected by these activities as well as by direct tax/transfer policies. Urban policies have long been recognized as containing significant redistributive elements. Ethnic and minority groups have frequently attempted to capture urban political control to obtain economic and social gains through expansion of the patronage system. Indeed, the fact that a high proportion of the city's revenues are financed by the property tax and intergovernmental transfers, the incidence of which may fall upon noncity residents, enhances the tendency toward urban governmental expansion as a means of income redistribution.

A third theory regarding the tendency of government to expand was developed in an explicitly urban institutional context and focuses upon uneven technological developments. Baumal (1976) argued that services are less likely to benefit from cost reductions attributable to technology than are goods produced by manufacturing or agricultural processes. For instance, the productivity of an assembly line worker may be improved by technological change, but it is less likely that a social worker can increase productivity by applications of technology. Consequently, Baumal projected that the costs of a "unit of service" will rise relative to the cost of manufactured or agricultural products. Because of the high service component of government activity, one can expect increasing proportions of the national income to be devoted to the public sector. Although there are opportunities of technological innovations in administrative services (Holzer, 1976), they have not been as significant as in manufacturing. In those cases where technology has been applied to service provision, for example, medical technology, it has been associated with higher costs and has frequently involved a major change in the nature of the product.

Urban governments provide a higher portion of what can roughly be called "service" activities than either the private sector or the federal government. In 1976, education, public welfare, health, and safety, all largely service activities, constituted 47% of expenditures for all cities, but 62% of expenditures for cities of over one million (U.S. Bureau of the Census, 1977). Thus, the explanation focusing upon different rates of technological change may have special salience in the case of urban public growth.

These three theories of government expansion are complementary since they concentrate on the different public sector functions. The public sector has indeed increased the scope of its various activities, including direct allocation of goods, redistribution of income, and regulation. As valid as these theories of government growth might be, it can no longer be taken for granted that the public sector will continue to expand at rates similar to those of past decades. Indeed, the predictive ability of social science theories is constrained by the often implicit assumption that trends in the future will be as they were in the past, *if* trends continue in the future as they have been in the past. But as discussed in the following sections, some basic societal nonerratic trends seem to be emerging that may dampen the rate of government expansion.

PORTENTS OF RETRENCHMENT

Governmental growth can no longer be assumed. In fact, the continuation of previous rates of increase in government is incompatable with the maintenance of a mixed economy. This contradiction is one reason to suspect the rate of government expansion may decrease. Attempts to slow the growth of the public sector have ranged from legislative and executive branch policy initiatives, such as sunset laws and zero-base budget proposals, to direct referendum initiatives in states and municipalities. Clark (1978), for example, has shown that statutory spending limits have been enacted or proposed in nineteen states other than California, and attitudes also appear to center on tax reductions, even outside of California. Relatedly, recent public opinion polls indicate that a substantial proportion of Americans oppose tax increases even if it would cause a contraction in public services. When asked about decreasing services or raising taxes—"which would you prefer?"—73% of the non-California respondents who favored Proposition 13 chose services reductions, compared to 12% favoring other tax increases. Even among nonsupporters of the tax reduction amendment, 52% favored service reductions compared to only 33% favoring other tax increases (CBS/New York Times, 1978). This finding contrasts with those reported in the early 1970s. The latter showed Americans to be fairly evenly split between opting for smaller government with fewer services and larger government with expanded services (CBS/New York Times, 1978). Furthermore, there is a significant increase in the frequency and intensity with which a contraction position is being advocated, as evident by the number of grass roots initiatives throughout the country as well as the policy stands articulated by decision-makers. One critical factor for explaining these changes is the present and the projected state of the national economy.

THE ECONOMY: CONDITIONS AND PROJECTIONS

Expectations regarding the future state of the economy are contributing to the reevaluation of government's role in the nation's domestic affairs. The 1960s, a period in which much social reform legislation was introduced, was also a period of steady economic growth. The 1970s by contrast has been a period of sluggish economic growth. "Stagflation" is a term used to describe the combination of two major recessions (the most recent of which was the sharpest downturn since the great depression) with inflation. As a result of the poor economic performance, economic expectations have been revised downward. A recent survey revealed that only 9% of the population felt their income would rise more than prices over the next twelve months (Gallup, 1978). Lowered economic expectations have led to greater caution regarding expansion of government programs, because the opportunity cost in terms of forgone private consumption is likely to be greater in a period of slow or no economic growth.

The public's expectations are congruent with systematic analyses and projections. A summary of congressional hearings into the prospects of the economy was titled, *U.S. Long Term Economic Growth Prospects: Entering a New Era* (1978). The concensus among those who testified in the hearings was that the nation has entered a period of slower economic growth. The most frequently voiced prospect was for long-term slower economic growth. The transition to slower economic growth will inevitably affect all phases of the economy, including rates of investment, technical progress, resource utilization, and the role of government. Furthermore, the congressional summary cited evidence, in counterpoint, that preference shifts away from consumption/ acquisition behavior were afoot that might mitigate the social tensions resulting from economic change. These socioeconomic preference shifts are directly influencing the formation of urban policy. The report concluded that slower economic growth will embody many potential sources of social stress and will strongly challenge future economic policy. Slower growth, however, need not be interpreted as economic stagnation, if these challenges are met with adequate planning and innovative responses.

Perhaps the most important proposition that underlines the changing economic performance is that economic development is inescapably an entropy-creating process (Georgescu-Roegen, 1971). Central to the entropy law is the principle that nothing (in a closed system) can grow forever. A limiting factor to economic growth is the diminution of the environmental carrying capacity in the form of the depletion of energy and mineral resources and pollution. Scarcity of resources adversely affects economic growth.

There are, of course, also social limits to economic growth. Harman and Thomas (1976: 50), for example, identified seven social factors associated with the modern economy that may result in diminished performance as the economy continues its present course of performance:

(1) Increased interdependence results in vulnerability to disruption and rigidity.
(2) Larger scale leads to decreased access to leaders, resulting in poor information flow, impersonality, and diminished significance of the individual.
(3) Complexity encourages increased reliance upon elite experts and diminishes the systems understanding by individuals.
(4) Formal controls, increasingly forcible and broadly cast, replace custom and internalized controls.
(5) Specialization hinders normal political checks, and renders individuals more vulnerable to economic change.
(6) There is a tendency toward low performance due to subsystem breakdowns, and inertia.
(7) The emergence of institutional boundaries and turf rivalries make planning more difficult.

Although the social limits to growth are less tangible than resource limits, the frustrations that result from social complexity may be greater than those caused by physical limitations, precisely because they are less tangible and therefore more difficult to understand. Each of the seven social limits to growth is associated with management difficulties not only for the economy, but for the entire social system. Management difficulties call for more coordination and new social strategies. Yet, cutting social programs might appeal more to the general public than restructuring the service delivery system.

The impact of changing economic circumstances is of special importance to the future of cities whose economic base has been deteriorating as jobs have shifted first to the suburbs and more recently to non-metropolitan areas. These regional shifts are part of the same factors that are causing slower growth nationally (Blair, Masachio, 1978).

The changing perceptions of the future economy coupled with frustrations with its sluggish performance, have been important factors in the reevaluation of the role of government in public policy. It is noteworthy to contrast these with earlier perspectives. Galbraith (1958: 198), in an influential work proposed that a "social imbalance" existed between the public and private sectors:

> The final problem of the productive society is what it produces. This manifests itself in an implacable tendency to provide an opulent supply some things and a niggardly yield of others. The disparity carries to the point where it is a cause of social discomfort and social unhealth. The line which divides our area of wealth from our area of poverty is roughly that which divides privately produced and marketed goods and services from publicly rendered services. Our wealth in the first is not only in startling contrast with the meagerness of the latter, but our wealth in privately produced goods is, to a marked degree, the cause of the crisis in the supply of public services. For we have failed to see the importance, indeed the urgent need, of maintaining a balance between the two.

The broad policy implications of this position were that more resources should be devoted to public consumption. Galbraith's argument was advanced at a time when private consumption had grown at rapid rates and was at historic peaks. Many middle-class Americans viewed their consumption levels as almost or soon to be luxurious. Certainly, consumption exceeded the prewar expectations of most Americans. While never explicit, the general expectation was that the "social imbalance" could be corrected merely by allocating a larger share of the growth increment to government activity. During the 1960s substantial growth in both the public and private sector was possible simultaneously. But when, as currently, significant expansion of the public sector can occur only by cutbacks in private consumption levels to which most Americans have become accustomed, the case for expansion of government activities elicits a critical reaction.

The "social imbalance" argument focused upon production rather than distribution and redistribution. It primarily addressed the question "what kinds of goods should be produced?" rather than how to distribute and/or redistribute them. Other writers of the period argued for greater equality in the distribution of income, or for a War on Poverty (Harrington, 1963).

The idea that economic growth is critical for raising the standard of living among the poor has systematically been advanced by Kahn and associates (1976). But even earlier, most of the "freedom budgets" resulting from the civil rights movement demands were based upon the idea that inequality could be reduced by allocating to the poor some portion of the growth increment. However, policy commitments to greater equality and antipoverty programs will be hindered by the slow growth prospects of the future economy. The "fixed pie" perspective will make redistributive issues increasingly difficult to negotiate. If gains by some groups result in relatively paralled declines to other, the prospective of a politics of "selfish redistribution" casts a vision of what might be termed a Hobbesian future of conflict politics. Although the "crisis" issue of unemployment, food costs, and energy have levied the greatest hardships on the poor, they have had also a significant impact on middle-income families, whose expectations for increased prosperity have become sensitive to redistributive policies. Furthermore, the same difficulties encountering redistributive policies in a slow-growth economy are present with other government programs designed to reduce inequality without explicit transfers, such as affirmative action or educational opportunities.

EVALUATION, SOCIAL SYSTEM, AND URBAN NEEDS

Concurrently with the economic strains, dissatisfaction with government intervention has increased, and the retrenchment advocacy has been gaining ground. Dissatisfaction with the results of government social programs is widespread not only among the beneficiaries but also among those responsible for delivery as well as social and academic critics. One government official articulated this widely held view as follows:

> The federal government has developed two defects that are central to its existence: (a) it does not know how to tell whether many of the things it does are worth doing at all, and (b) whenever it does decide something is worth doing, it does not know how to create and carry out a program capable of achieving the results it seeks. [Carlson, 1969: 75]

In past decades the principal criticism against government intervention was articulated in terms of a threat to individual liberty. Hayek

(1949), for example, took the position that government activity must be strictly circumscribed if Americans are not to march on the "road to serfdom." More recently, Milton Friedman conjectured that the expansion of government size limits individuals' free choice. The private enterprise economic system was advocated on the basis of the sociopolitical value of freedom. In Friedman's (1962: 12) words, "As liberals, (in the classic sense) we take freedom of the individual, or perhaps the family as our ultimate goal in judging social arrangements."

Interventionist liberals responded to such arguments along the lines of Anatole France: "what freedom for whom?" Critics of government intervention argued for the efficiency of market allocation by fitting social reality into the competitive model of economic interaction; that is, although government interventionist programs could help some groups, they were inefficient for the society at large. The standard of judgment was normally the optimum under perfect competition, rather than an attainable social state. Boulding (1969: 5) referred to the fairy-tale nature of such policy evaluation standards as "a stirring intellectual drama which might be subtitled "Snow White (the fairest of all) and the Seven Marginal Conditions!"

Contemporary critics have modified the traditional noninterventionist argument in several important respects. Effectiveness and efficiency are postulated to be the society's primary goals. Furthermore, effectiveness is not evaluated with respect to an idealized state, e.g., a social welfare function. Rather, comparisons are made projecting the state of society with or without a particular government policy. Significantly, such analyses extend beyond the narrowly defined economy into noneconomic aspects as well. Thus, it is conjectured that government policies, in particular those targeted at social problems, contribute to the long-run decline in economic growth. Roberts (1978), among others, has suggested that the Keynesian paradigm of economic policy has "broken down" because of its inadequacy in dealing with problems of supply. The Keynesian model, the argument goes, is structured to generate policies that stimulate demand. But in a resource-scarce environment, increasing demand may not generate sufficient additional output, and will produce inflationary pressures. Similarly, the lack of incentives reduces supply: as tax rates increase, the incentive to earn taxable income diminishes. Consequently, individuals will work less, production will decline, and the reduced supply will enhance inflationary pressure. Urban programs designed to ease the impact of unemployment likewise tend to decrease aggregate supply, according to this view. Fieldstein (1973), for instance, showed that there is little difference between the after-tax income of the average hourly worker and what he or she would receive from untaxed unemployment compensation.

Corporate tax rates have also a supply-side effect. High tax rates reduce corporate investment by making after-tax rates of return less attractive. For example, if a 10% after-tax rate of return is necessary for an investment in capital plant or equipment to be undertaken, and if the corporate tax rate is, say, 50%, then the before-tax profit rate must be 20%. Thus, critics of "big government" maintain that higher tax rates have lowered investment, the latter generating smaller productivity increases, and no/slow economic growth (Burns, 1976: 10).

The proposition that continued economic growth will become increasingly difficult to achieve due to resource scarcity and societal complexity is logically consistent with (although not deducible from) the argument that high levels of government activity hinder economic growth. These factors may in fact operate simultaneously. However, different policy implications can be derived from the two positions. Advocates of the former position might suggest credit controls and regulations to insure an equitable distribution of income. But such policies could aggravate the problem from the perspective of those who believe the difficulties emerge from a social system too complex to be amenable to successful policy intervention. Obviously, the resolution of the policy issue, how to deal with slow economic growth, has implications regarding the distribution of wealth so that parties are seldom disinterested in the outcome. Kahn (1976) has suggested that the need of limit growth, and the implications for economic regulation are often attempts by middle class individuals to secure their economic positions.

Noninterventionists contend that government programs are inefficient, but not in the traditional sense that they fail to attain an idealized optimum. Rather, they frequently point to well-intended government programs that have, in the long run, unintended consequences detrimental to causes and groups favored by the traditional urban liberal. Federal Housing Administration programs, for example, have been characterized as adverse to cities, because they have encouraged development in the suburbs at the cost of deteriorating the real estate and the tax base of the city. Freeway programs designed to increase access to certain parts of the city have, in retrospect, contributed to the breakup of older neighborhoods. The housing demolitions necessary for new construction have diminished the tax base not only by the value of the houses directly removed, but also by the value loss of the remaining properties bordering the newly created highway. Similarly, it has been suggested that welfare programs do not solve urban problems but only attract more poor to the city. The judicially imposed policy of school desegregation further illustrates the nature of the unintended consequences argument. The purpose cited by the policy advocates was to desegregate public schools, improve minority students' education, and perhaps encourage social integration and promote racial harmony.[1] Critics of desegregation programs contend that such programs produced the opposite effect by encouraging white flight from the cities among affluent white families. Furthermore, it is suggested that students who are bussed are as socially isolated as they would have been in segregated schools. Thus, school desegregation has been alleged to exert the adverse impact of contributing to the racial polarization of the polity. In a more general vein, Allman (1978: 46) contends that the "great society" program "helped set up cities for the crisis that broke a few years later."

The tenacity of such criticisms goes beyond the documentation of partial or total failures of specific urban programs. It should not be surprising that some government social programs produce unintended results; these, however, are inherent in the nature of the complex urban system. Urban society is a complex system, and complex systems are by nature *counterintuitive* (Forrester, 1969). Because of the complicated and unpredictable patterns of interdependence among the various subsystems, policies attempting to affect the level of a target variable set in motion "self-correcting" or "equilibrating" forces that tend to reestablish the variable at the prepolicy level; the side-effects might make matters worse. As Forrester (1969: 9) puts it:

> It has become clear that complex systems are counter intuitive, that is, they give indications that suggest corrective action which will often be ineffective or even adverse in its results. Very often one finds that the policies that have been adopted for correcting a difficulty are actually intensifying it rather than producing a solution. . . .
>
> The intuitive process will select the wrong solution much more often than not. A complex system—a class to which a corporation, a city, an economy, or a government belongs—behaves in many ways quite the opposite of the simple systems from which we have gained our experience.

Because of our experience with relatively simple systems we also tend to envision noncomplicated relationships in complex systems and fail to understand higher-order consequences. Indeed, the reasoning developed by Forrester and others parallels the traditional economic-efficiency argument, but is more encompassing. The standard of comparison is not the idealized model of competition, but the world as it could exist within the model. Contemporary critics are more likely to use the "unheavenly city" in the "unheavenly world" as their standard of comparison.

A profound policy dilemma is implied by the counterintuitive nature of public policy because policies that are indirect or roundabout take a longer time to work themselves through complex systems. This, however, is unsatisfactory to public policy-makers desiring immediate gratification, and to social moralists who want to solve problems quickly. In other words, policies that have the greatest intuitive and political appeal are likely to be least successful.

Oates and his associates (1969) developed a rather simple, two-equation model of the city that can be modified to represent the self-equilibriating argument.

$$W(t + 1) = a - b\, D_t \quad [1]$$
$$D_t = y - z\, W_t \quad [2]$$

where W_t = an index of urban welfare (such as income in period t, and D_t = an index of urban deterioration (such as below code housing) in period t. The first equation states that the greater deterioration in any period, the smaller will be the urban welfare index. The second equation postulates that the level of deterioration is determined by the level of urban welfare in the current period. The measures D and W represent the urban policy problem: how to increase W or alternatively decrease D? This system of equations is in equilibrium when substituting equation [2] into [1] and rearranging:

$$W(t + 1) = a - b\,(y - zW_t) \text{ and specifying the}$$
$$\text{equilibrium condition: } W_t = W(t + 1) = W_e: \quad [3]$$
$$W_e = (a - by) \div (1 - bz). \quad [4]$$

The intuitive approach to the problem of raising the urban welfare index might be to increase W directly, say, by income grants. Alternatively, a direct policy attempt to decrease deterioration might take the form of financing a store-front fix-up program. Although such interventionist programs might increase the level of welfare temporarily, they will not affect the longterm dynamic process.[2] Only policies that affect the parameters of the equation—a,b,y, and z—will permanently improve urban welfare. This relationship underlies the aphorism that "problems can't be solved by funding them away." According to this system structure, once the stream of direct money subsidy dries, the old equilibrium will be reestablished.

Another aspect of the self-equilibrating nature of a complex system was developed by Tullock (1974). Government programs create temporary gains for some groups, but once the gains become embedded in the incentive system, the benefits are discounted and thus lose their potency. For instance, a program might be introduced to improve the working conditions of industrial workers. The program would attract more individuals into the occupation, wages might fall, and consequently, the employees would not be better off as a result of the program. On the other hand, if wages were rigid, competition for the improved jobs would intensify, and qualifications would increase. In order to get the better job, individuals would be willing to expend extra resources, until the marginal benefits flowing from the improved working conditions would equal the costs (monetary, social, and psychological) of qualifying. Whereas the policy change would have helped those who already had a job in the affected industry, newcomers would not benefit; the latter would have paid for the improvements in the form of higher entry costs. Their gains would only be transitional. The problem with such form of government intervention is that once successfully implemented, repeal will cause real losses to individuals who made plans based upon the program. For example, the removal of the federal mortgage interest and property tax deductions would create hardships for those who bought homes—at higher prices than otherwise—based upon the tax subsidy.

Typically, the urban liberal response to such criticism is that the real problem is not government intervention per se, but poorly conceived and inadequately implemented government programs. Furthermore, government is conceptualized as a learning system; its future policies will avoid the errors embodied in earlier programs, thus achieving a higher success rate. From this perspective, the "big government" critique does not establish a case against government intervention, but rather calls for the development of better conceived, more realistic and implementable programs. Indeed, policy-makers can enhance urban welfare directly and indirectly through policies targeted at the system's underlying parameters. But it is precisely the parameters and the target variables of the urban system that policy-makers and urban liberals have trouble delineating and defining: "The parameters and structural changes to which a system is sensitive are usually not self-evident. They must be discovered through careful examination of system dynamics" (Forrester, 1969: 111).

A related criticism of government intervention pertains to the government's capability to manipulate parameters of the urban system, once these become analytically comprehensible. Such parameters, according to some observers, are beyond the purview of state action. In Banfield's (1970: 257) words,

> Powerful accidental (by which is meant non-governmental and, more generally, non-organizational) forces are at work that tend to alleviate and even to eliminate the problems. Hard as it may be for a nation of inveterate problem solvers to believe, social problems sometimes disappear in the normal course of events.

Moreover, "if there *are* solutions to these problems they are not *governmental* ones, which is to say that one cannot implement them by calling into play the state's ultimate monopoly on the use of force" (Banfield, 1970: 257). Economic growth, demographic change, and the process of "middle- and upper-classification" are, according to Banfield, the key parameters that will improve the conditions of the "Unheavenly City." Things may get better, but whether or not they do is beyond government will and policy. Forrester expressed a similar view regarding both social and economic programs.

> We have seen such futile pursuit of ineffective policies in the last two decades in government efforts to solve the urban problem. In a similar way, dependence on low leverage policies probably explains the present frustration in dealing with economic situations. [Forrester, 1978: 383]

Similar policy implications may be derived from Jenk's (1972) argument that family characteristics rather than the quality of the school determines academic success and future economic well-being. Likewise, the findings that the distribution of income is not significantly affected by income policies has led to the contention that ability, effort, intelligence and the opportunity structure rather than government policies determine economic well-being.

Critics of intervention have focused more upon urban policy as implemented at both the federal and local level than other types of government programs although the arguments have been generalized to other activities. The urban policy emphasis is evident from the titles of the two most influential books of this school: Banfield's *The Unheavenly City* and Forrester's *Urban Dynamics*. The urban focus of anti-interventionist analysis probably results from the greater complexity of urban society compared with smaller-scale subunits of the polity.

The view that complex systems are difficult to manage and inherently ungovernable underlies a more radical group of social critics who advocate a thorough reorganization of government to simplify the system. Only in smaller subsystems will government interventionist policies work, and at the same time, the need for government intervention will decrease. Although not necessarily opposed to government intervention, these critiques advocate "small government"; the city, state, and in particular the polity are too large to be amenable to effective management and responsive governance.

Schumacher (1973) was perhaps foremost among those advocating reorganization as prerequisite for effective and responsive government. He linked inefficient large-scale technology with complex social organizations, claiming that regional organization and the geographic distribution of population will be the critical issues in the future. System reorganization proponents maintain that large-scale society is not only difficult to manage, but also lacks mechanisms or opportunities for social integration; it lacks "communitas." As Bookchin (1974: 146-147) put it, "The larger cities of the world are breaking down under sheer excess of size and growth. They are disintegrating administratively, institutionally, and logistically." Accordingly, more of the burden of collective action ought to be lodged within neighborhoods where people know "what is going on" (Norris and Hess, 1975). The perception of the city as a complex system does not rule out possible government intervention; it calls, however, for more careful analyses and evaluations grounded on more realistic expectations.

DIMENSIONS OF URBAN NEEDS

Another strand of the position that government ought to contract its activities is the argument that, in general, social and economic conditions have greatly improved. This observation is related to the previous one because it is suggested that the factors generating improvement in social welfare are embedded in the dynamics of the social system and unrelated to government policy. Nonprogrammatic impacts of government spending may even wash out targeted programs.

Banfield (1968: 21) at one extreme, suggested that urban problems are programmed to keep ahead of performance. If real problems do not exist government employee's anxious to justify their employment or expand their authority may create "crises." More recently, Moynihan (1978) suggested that if a bureaucracy were developed to study and respond to problems caused by the sunbelt shift of industry, it might become a lobbying group interested in "artificially" perpetuating the crisis. Likewise, social workers, criminologists, and teachers can be argued to be interested in perpetuating "crisis" images. It has frequently been pointed out that it is impossible to eliminate the lower 20% of the income hierarchy, yet welfare programs always appear to be concerned with this inequality.

The contention that, in general, things are getting better is strongly supported by long-run trends pertaining to indicators such as income, health, and housing.[3] Median family income has risen over 88% since 1947; the number of people with income below the official poverty line has fallen from 32% of the population in 1947 to 12% in 1976, and the proportion is even smaller if in-kind transfers are included in the estimates. Almost all of the aggregate health care indicators suggest improvement: a higher percentage of the population has access to hospitals, life expectancies have increased, and infant mortality rates have declined dramatically. Using measures such as housing units lacking some plumbing facilities or the number of persons per room the housing stock has greatly improved in the past thirty years.

Other conditions, however, have not improved. Crime rates, for instance, have continued to increase since 1947 although the prison population rate has remained rather constant and the crime rate has actually declined in some cities very recently. Levels of educational achievement have not improved appreciably in recent years although educational expenditures have increased. Income inequality has been scantly altered although absolute income levels have increased. Recent studies of subjective life quality also indicate a rough stability over time (Campbell et al., 1976).

Table 1 lists major life quality indicators according to whether they have improved, worsened, or remained unchanged. In general, measures

TABLE 1: TRENDS IN URBAN INDICATORS

Conditions Improving	Conditions Unchanged	Conditions Worsening
Absolute income	Income distribution	Governability
Health indicators	Racial desegregation	Crime
Absolute poverty	Subjective happiness	Pollution
Housing conditions	Relative poverty	Traffic congestion
Years of schooling	Unemployment	Neighborhood quality

of social status that can be affected by economic growth or technology show improvement. But conditions that are affected by social behavior evidence little progress. The latter are, according to critics of government intervention, the most impervious to government manipulation.

RETRENCHMENT IN CITIES

Pressures for further government expansion and counterpressures for retrenchment are evident in cities, to a greater degree than elsewhere. The present section examines some possible reactions that could develop in an era of retrenchment.

The mounting pressures for government cutbacks already exert a severe impact on urban residents. The older cities of the Northeast and the Midwest are now, more than ever, experiencing a period of sustained economic and social decline, accompanied by a crisis of political authority and a gloomy fiscal future. The cities' dependence on the federal government is manifested in their revenues. Buffalo, St. Louis, Cleveland, Detroit, Philadelphia, and Baltimore, for example, all have received at least half of their general revenues from the federal government. Federal cutbacks will increase the cities' budget deficits, and will inevitably lead to contraction in basic services, layoffs of city employees, and worsening of the social-economic conditions of those who depend on the governments' various social programs. For example, Boston Mayor Kevin White projects the layoff of 1,000 city employees this year. Philadelphia, which fired 900 city workers last year, will lay off more employees to avoid a combined city and school-system deficit of $165 million by 1980. Newark has begun firing some 440 city workers, including 200 from an already reduced police force. Cleveland defaulted on $14 million in notes to local banks—the first big city to default since the Depression. The projected cumulative budget deficit for New York City is $2.3 billion by 1982, and the city is experiencing heavy cuts in education and health care, and some 60,000 jobs have already been cut from the city payroll. Overall, as Table 2 shows, city government employment (one indicator of retrenchment) leveled in 1974, although the general trend was still somewhat positive. A decline in 1976 accentuated this slower or negative growth trend. Payrolls also increased, although the increase in 1976 was well below the percentage of increase in general process.

Relatedly, Congress seems to be taking a more conservative fiscal stance. It did not renew a program of antirecession fiscal assistance, known as countercyclical aid, that had brought $3 billion to cities with unemployment rates above 6%; it cut back the public service jobs financed by the Comprehensive Employment and Training Act, and it did not pass a $400 million program of grants to states to encourage urban revitalization. The executive branch's urban policy seems to be formulated under the projected image of the "Lean Society." As President Carter told the National League of Cities in November 1978, "It will be an austere budget. However, I promise that the cities will bear no more and no less than a fair share of budget restraints."

Cities are experiencing the new reality of "rising needs-declining revenues." They are undergoing a crisis because the rising expectations and service needs of city dwellers grow at much faster rates than local revenues. More explicitly, the changing demographic and social composition of big cities necessitates far greater investments in infrastructure and services, including public welfare, special educational programs, more intensive policing, and more elaborate health, sanitation, and sewerage systems, if the city is to maintain itself. Furthermore, the maintenance of the civil order depends to a large extent on the provision of these services. One consequence of the urban violence of the 1960s was the activation of social and economic programs, many sponsored by the federal government, targeted primarily at low-income urban minorities: the War on Poverty. The Elementary and Secondary Education Act, Model Cities Program, Manpower Development and Training Act, Headstart, medicaid, revised FHA mortgage guidelines facilitating central-city investment, and affirmative action programs. Low-income minorities, the elderly, and the poor depend on these programs and expect government to continue and even expand them.

Wherever big cities have expanded their scope of activities, their local revenues have not increased at the same rate, particularly since

TABLE 2: CITY GOVERNMENT EMPLOYMENT AND PAYROLLS: 1950 TO 1976 (for October 1967 and 1972 based on complete count of all cities; other years based on sample and subject to sampling variation)

Year	All Employees, Full-Time and Part-Time (1,000)		October Payroll (mil.)		Average Annual Percentage Increase		Full-Time Equivalent Employment (1,000)			Average Earnings in October, Full-Time Employees	
	Total	Excl. Education	Total	Excl. Education	All Employees	Monthly Payroll	Total[a]	Education	Other	Education	Other
1950	1,311	1,106	290	230	23.2	8.9	NA	NA	NA	NA	NA
1955	1,436	1,238	414	337	1.8	7.4	1,262	182	1,080	$422	$315
1960	1,692	1,439	583	471	3.3	7.1	1,447	225	1,222	502	387
1965	1,884	1,560	818	649	2.2	7.0	1,638	282	1,356	603	480
1967	1,993	1,633	972	769	2.9	9.0	1,715	306	1,410	664	546
1970	2,244	1,815	1,361	1,062	4.0	11.9	1,922	359	1,563	838	681
1971	2,273	1,838	1,482	1,167	1.3	8.9	1,960	366	1,594	876	735
1972	2,375	1,918	1,654	1,302	4.5	11.6	2,029	378	1,650	951	792
1973	2,471	1,992	1,855	1,441	4.0	12.2	2,109	402	1,707	1,405	846
1974	2,491	2,009	1,985	1,560	.8	7.0	2,127	405	1,722	1,060	909
1975	2,506	2,074	2,129	1,725	.6	8.3	2,142	376	1,782	1,130	972
1976	2,443	2,021	2,235	1,804	−2.5	5.0	2,107	360	1,747	1,207	1,036

NA: Not available.
[a]Includes only those school systems operated as part of the general city government.
[b]Change from 1946.
SOURCE: U.S. Bureau of the Census, *City Employment*, annual.

business and the better-off taxpaying residents have been leaving the cities. To offset this imbalance, municipal revenues have been increased by enlarging the federal and state aid to the cities. Furthermore, because of the increased fiscal dependence, in the future big cities will be even more directly dependent on national and state political institutions and policy decisions made at these levels. However, national and state priorities are changing at the same time as city tax bases are declining.

Although city "needs" cannot be deduced rigorously from expenditure patterns, the latter may serve as one indicator. The larger scope of big-city governments compared to small local governments is apparent from Table 3. General expenditures are $844 per capita for cities of over 1,000,000 population as compared to $359 per capita for all cities. The larger expenditure of cities is linked to larger revenue collections. Because of the initial high levels of taxation and of the sources of revenue, maintenance and expansion of urban governmental services may be more difficult to attain than in areas with a smaller initial tax burden. Property taxes, the target of much of the taxpayer dissatisfaction, are twice as large in the major cities as in the average city. Dissatisfaction with the property tax is evident in recent opinion polls. A study by the Advisory Commission on Intergovernmental Relations showed that the property tax is considered to be the least fair tax, although the federal income tax is not far behind. In 1978, 32% of the surveyed individuals felt that the local property tax was the least fair (ACIR, 1978). The dissatisfaction with the property tax, coupled with the already high level of property taxation in major cities leads to the projection that increased government activity (if any) is unlikely to be financed from that source.

Table 3 also indicates the importance of intergovernmental revenues as a source of urban finance. Nearly half of large-city finances come from intergovernmental revenues. But reliance upon such revenues as a base for maintenance and/or expansion of urban programs is no longer guaranteed; federal and state governments are also under pressures to reduce spending and taxes. Intergovernmental revenues are likely targets for cutbacks partly because intergovernment transfer programs seldom require large bureaucracies within the granting agency. Thus, there is less internal support for intergovernmental aid programs than many others. Furthermore, problems created by decreases in intergovernmental aid are more likely to be faced by the grantee government than the granting unit. Preliminary evidence regarding federal budget

TABLE 3: CITY GOVERNMENT FINANCES (city governments, finances by population, size groups: 1975)

Item	All Cities	Less than 50,000	50,000 to 99,000	100,000 to 199,999	200,000 to 299,999	300,000 to 499,999	400,000 to 999,999	1,000,000 to More
Per Capita[1] (dollars)								
General revenue[1]	367	196	286	336	393	413	520	922
Intergovernmental	145	63	91	119	154	156	217	438
From state governments	96	37	54	69	81	81	105	363
From federal governments	43	22	33	43	61	65	101	70
Revenue sharing	16	12	15	17	20	20	21	26
From local governments	6	4	3	7	11	10	11	5
Taxes[1]	156	87	136	152	159	161	213	378
Property	96	57	101	111	96	91	121	198
Sales and gross receipts	20	10	17	17	17	20	29	56
Current charges	10	29	37	40	46	56	55	64
Water supply and other utility	61	54	46	69	50	44	77	87
General expenditure[1]	359	194	296	341	395	414	532	844
Education	53	18	47	59	57	67	72	154
Highways	28	27	30	28	31	26	36	26
Public welfare	28	1	3	7	16	4	46	163
Health and hospitals	27	9	13	14	18	18	48	102
Police protection	39	25	33	38	40	48	58	73
Fire protection	21	14	24	28	28	21	30	29
Sewerage and sanitation	38	32	31	39	44	51	48	53
Housing and urban renewal	13	4	11	13	21	21	15	40
Interest on general debt	17	8	11	14	23	22	21	49

probabilities is particularly ominous. Both direct aid to cities and indirect aid through federal programs appear to be caught in a squeeze between the desires to reduce the federal surplus, avoid a tax increase, and maintain or increase defense appropriations. But increases in the defense budget might only indirectly exert a positive impact on urban areas because of the geographic distribution of military spending. In general, cities pay more in taxes for defense programs than are spent directly in the areas for defense procurements.

Pressures on city finances are not, of course, new. In the past urban policy-makers were compelled to meet a variety of objectives that aggravated the current budgetary situation. Many cities postponed meeting certain long-term obligations in order to finance their services and social programs. Cities have been only partially funding pension obligations. For example, it has been estimated that Pittsburgh would be required to triple its current pension payments, increasing total expenditures by over 10%, in order to fund fully the current pension obligations plus amortization over forty years of past service liabilities (Peterson, 1978: 86).

Postponement of urban infrastructure maintenance has been another typical response to fiscal pressures that shifts the burden to the future. Peterson's (1978: 86) study indicated that "several of the older cities operating under fiscal pressure have had spectacular reductions in their capital spending." Consequently, local government spending on sewers, roads, public building, and the like, will soon have to increase in many cities. Yet, added spending on infrastructure will create further pressures to cut urban social programs.

Population shifts, reflecting both a preference for nonurban living and a fundamental spatial economic shift, constitute a dynamic force that will accentuate demands for urban cutbacks. Between 1970-1975, over half of the cities with a 1970 population of over 100,000 experienced net declines in population. This, in turn, decreased the property tax base, causing total assessed values to fall in many areas. In other cities inflationary factors and rapid reassessment increased total assessed value, although seldom enough to cause the tax rates per dollar of market value to decrease.

As city population declines, costs will decline, but at slower rates given that much of a city's infrastructure is a fixed cost with regard to population. For instance, road repair expenditures are a function of number of miles of roads, vintage, and previous maintenance rather than population size. But even with regard to municipal employment it is difficult to affect cutbacks proportionate to population. Muller (1977) found that the number of personnel providing police, fire, and sanitation services increased in expanding cities at almost twice the rate of population increase. On the other hand, in declining cities population loss was four times the rate of public employment decline. Consequently, costs per capita will increase in cities losing population. Muller's findings not only indicate that cutbacks in declining cities are difficult, but also demonstrate that population redistribution is itself costly. The observation that costs per resident increase as population declines, does not necessarily imply that taxes paid by residents have increased, because in many cases intergovernmental revenues or payments by nonresident property owners have had an ameliorating effect. However, if they are necessary, higher property taxes will create a "development drag" regardless of the location of the current owner because, in the long run, they contribute to higher rents and business operating expenses.

A second reason that population loss adversely affects city revenues is that in/out city migration has not been selective. The relatively well-to-do leave the cities; the economically dependent, the poor, and marginally productive residents remain. Although the "back to the cities" movement may in the long run offset this trend, the effects of years of suburbanization among the more affluent will not be altered quickly.

The higher tax rates, greater dependence upon intergovernmental support, deferred physical maintenance, and population loss suggest that expansion of urban social programs will be slow at best, and possibly negative. Yet, urban social needs, and particularly those that have not changed in recent decades are disproportionally concentrated in cities. Furthermore, urban centers have greater difficulties in diffusing social problems associated with inequality because urbanization makes private granting a less effective strategy for ameliorating social problems (Blair et al., 1975: 498). As Muller (1977: 121) has demonstrated costs per capita in large cities are higher for three major reasons: (1) Because of higher density it takes more people to provide municipal services, such as police protection; (2) city employees receive higher wages; and (3) service needs of dependent populations are greater.

The new reality of "rising needs-declining revenues" presents difficult policy choices with respect to these factors. Cuts in budgets will inevitably lead to cuts in basic services, including police protection, health, education, and transportation. These will not only aggravate the symptoms of the urban problems, but may also induce more middle-income families to leave the city, which, in turn, will result in further reductions in revenues, more cutbacks, and so on. Cities are caught in the vicious circle of rising needs-increasing revenues, and cutbacks from either the public or the private sector will be immediately reflected in the operation of this cycle.

Pressures for reductions in the scope of government involvement in urban affairs are clearly significant, but "pressures" will not necessarily result in change because forces preventing cutbacks of government are also operating. Many of the same factors that were associated with the evolution of a large governmental sector will continue to operate. Two such factors in particular should be discussed: the potential for civil disturbances and bureaucratic resistance.

High density, complex, and interdependent society renders urban areas vulnerable to civil disturbance. In addition to the genuine desire to satisfy social needs, many of the Great Society programs (or, as some were initially called, "demonstration cities" projects) were initiated and expanded in response to the riots of the 1960s. While some commentators have argued that the riots were not born out of social protest but were rather for fun and profit, many mayors and leaders of poverty groups espoused opposite views. Thus, the risk of civil disturbance is one of the costs of retrenchment.

The potential for civil disturbances resulting from future retrenchment is compounded by the fact that many individuals from economically disenfranchised groups depend upon social programs to create administrative and service delivery jobs. Such employment opportunities offer a new path into the mainstream labor market. The importance of jobs created for the service providers should not be minimized in the analysis of the adverse consequences of retrenchment.

Of course, the riot response will not necessarily follow budget cutbacks in social programs. "Social learning" has undoubtedly occurred in the interim. Political inroads of minorities have opened many other possible reactions. Nevertheless, there have been numerous examples of civil distrubances in response to cutbacks, albeit mild by standards of the 1960s. Most major facility shut downs are associated with some form of labor and community protest. Clearly, a retrenchment policy will be affected by the expected extent and form of social protest (Gamson, 1965).

The nature of urban government organization will affect the budgetary outcome. Partly in response to the perceived need for social programs and partly because of the greater interdependence, urban bureaucracies have grown larger over time. Urban bureaucracies have gone through the process of unionization, making them strong pressure groups within the context of urban government and leading to higher wages. While the higher wage rates have contributed to the fiscal strain, union-management efforts to formalize roles and procedures will make cutbacks more difficult. Many cities, as we demonstrated earlier, have already experienced layoffs of employees. The interdependence that make cities vulnerable to civil disorders also make them vulnerable to public employee strikes that might occur in the event of large-scale layoffs. Unions also tend to advocate seniority as the major layoff criterion. Leven (1978: 321) has suggested that seniority cutbacks bear little relationship to efficiency or equity. In fact, because seniority is likely to be spread unevenly across various agencies, the use of seniority creates additional management problems and a tendency toward inefficiency. The cutback process could result in the replacement of lower seniority persons in new programs by less qualified but higher seniority persons who have been transferred from the area in which they have accumulated their experience. It has frequently been pointed out that the incidence of seniority cutbacks falls most heavily on minorities and women—groups that have only recently gained access to significant portions of the labor market. The problem is aggravated because these groups have looked toward previously expanding governmental employment opportunities as entry points into the mainstream economy.

THE URBAN POLICY RESEARCH AGENDA

The forces of retrenchment coupled with the reactions of those who have interests in continued rapid expansion constitute a challenge at the levels of policy analysis, formulation, implementation, and administration. Slower economic growth will require more careful evaluation of tradeoffs among policy goals because programs undertaken in one year are more likely to preclude other future programs since planners can be less confident that future expanded budgets provide resources for postponed programs. Tradeoffs currently are often only implicit. Institutional changes will be necessary to reduce resistance to changes so that expansion in some spheres can be matched by reduction in areas with less valued outcomes. A political economic model of churning rather than expansion will emerge.[4]

It is increasingly apparent that the interventionist critiques should be incorporated into urban policy formation. The criticisms regarding the lack of positive outcomes should be confronted by exact models of specifying intended or unintended outcomes. More complex views of social processes than those that supported many previous programs will be needed. Perhaps development of various "impact statements" is one step in this direction.

The rising needs-declining revenues vicious circle also presents cities with difficult policy choices. The considerations entering the urban policy-making process, the dilemmas facing policy-makers, and the impact of their decisions pose new and different research questions to urban analysts. As Secretary of Health, Education and Welfare, Joseph Califano put it:

> Today the liberals and progressives of our society must match their compassion and generosity with competance and efficiency. Unless we accept and meet the challenges of austerity with good management, we will surrender to an undiscriminating Proposition 13 mentality that will do violence to the concepts of social justice on which the programs of the New Deal and the Great Society are so soundly based.

Obviously, one does not have to take a liberal or a conservative view to see that sound management in an era of retrenchment will be a critical factor in years to come. But urban analysts lack the information and knowledge of cities' behavior under conditions of decline. The systematic study of program implementation and evaluation is the newest social science discipline, and findings pertaining to the effectiveness and efficiency of urban social programs are tentative at best, lacking in analytical sophistication and theoretical rigor (Nachmias, 1979). In fact, policy decisions based on the present state of knowledge might have adverse effects on the incremental gains achieved through a few programs. Under the constraints of a "Lean Society" students of urban affairs will be increasingly concerned with problems of conflict management, productivity, and program effectiveness and effiency. A better understanding of such problems might make city life more tolerable in an era of retrenchment.

NOTES

1. These reasons, of course, are not the judicial support for desegregation orders although they may have influenced the thinking of judges.

2. Figure 1 illustrates the temporary nature of interventionist policy as described by the Oats model. Line e is the long-run equilibrium as determined by the underlying parameters. A government program is designed to increase welfare directly in period t, and the program is terminated in period tc. Upon termination the index of welfare will revert toward the original equilibrium by tn.

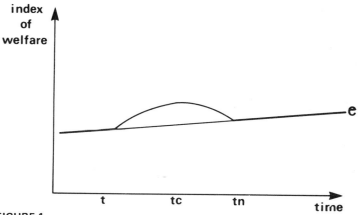

FIGURE 1.

3. For a detailed discussion of social indicator trends see "America in the seventies: some social indicators," *Annals* 435 (January 1978).

4. In order to illustrate the need to know more about public retrenchment, Glassberg (1978: 325) has pointed out that Widavsky's classic *Politics of the Budgetary Process* has three chapters directly dealing with the budgetary base: "Defending the base: guarding against *cuts* in the old programs"; "Increasing the base: inching ahead with existing programs"; and "Expanding the base: adding new programs." The chapter on "Defending the base begins by stating, "A major strategy in resisting cuts is to make them in such a way that they have to put back."

REFERENCES

ALLMAN, T. D. (1978) "The urban crisis leaves town and moves to the suburbs." Harper's (December): 41-56.

BANFIELD, E. C. (1970) The Unheavenly City. Boston: Little, Brown.
BAUMAL, W. T. (1967) "Macroeconomics of unbalanced growth: the anatomy of urban crisis." Amer. Econ. Rev. 57 (June): 415-426.
BEHN, R. D. (1978) "Closing a government facility." Public Admin. Rev. 38, 4: 332-388.
BLAIR, J., G. GAPPERT, and D. WARNER (1975) "Rethinking urban problems: inequality and grants economy," pp. 471-506 in G. Gappert and H. Rose (eds.) The Social Economy of Cities. Beverly Hills: Sage Publication.
BLAIR, J. and R. MASSACHIO (1978) "Regional shifts and post-industrial society," in Miller and Wolensky (eds.) The Small City and Regional Community. Stevens Point, N.J.: University of Stevens Point.
BOOKCHIN, M. (1974) The Limits of the City. New York: Harper and Row.
BOULDING, K. (1969) "Economics as a moral science." Amer. Econ. Rev. 69 (March): 1-12.
BURNS, A. (1976) "Interview," Challenge, (January/February).
CAMPBELL, A. (1978) Changing Public Attitudes on Government and Taxes. Washington, DC: Advisory Commission on Intergovernmental Relations.
--- P. E. CONVERSE, and W. L. RODGERS (1976) The Quality of American Life. New York: Russell Sage.
CBS/New York Times (1978) "CBS/New York Times Poll, Part I, June 1978," in Local Distress, State Surpluses, Proposition 13: Prelude to Fiscal Crisis or Opportunities? Hearings Before Subcommittee on the City of the Committee Banking Finance and Urban Affairs of the House and the Joint Economic Committee, 95th Cong. 2nd Sess., July 25-26.
COLMAN, J. S., S. D. KELLY, and T. A. MORE (1975) "Recent trends in school integration." Presented at the Annual Meeting of the American Education Association, Washington, DC, April 2.
CLARK, P. (1978) "What's happening outside of California?" Nation's Cities 15 (August): 26-27.
FIELDSTEIN, M. (1973). "The economics of the new unemployment." Public Interest (fall): 202-215.
FORRESTER, J. (1978) "A great depression ahead." Futurist 12, 6: 379-385.
--- (1969) Urban Dynamics. Cambridge, MA: MIT Press.
FRIEDMAN, M. (1962) Capitalism and Freedom. Chicago: University of Chicago Press.
GALLBRAITH, J. K. (1958). The Affluent Society. New York: Mentor.
Gallup Opinion Index (1978) August report #157, pp. 1-5.
GANSON, W. (1969) Power and Discontent. Homewood, IL: Dorsey.
GLASSBERG, A. (1978) "Organizational responses to municipal budget decreases." Public Admin. Rev. 38, (4): 325-332.
GEORGESCU-ROEGEN, N. (1971) The Entrophy Law and the Economic Process. Cambridge, MA: Harvard University Press.
HARMANN, W. W. and T. C. THOMAS (1976) "The challenges of noneconomic factors to economic growth," pp. 41-56 in U.S. Economic Growth From 1976 to 1986: Prospects, Problems, and Patterns, vol. 2. Washington, DC: U.S. Government Printing Office.
HARRINGTON, M. (1963) The Other America. Baltimore: Penguin.
HAYEK, F. (1949) The Road to Serfdom. Chicago: University of Chicago Press.

JAMES, O. C. and L. D. HOPPE [eds.] (1969) Urban Crisis in America. Washington, DC: Washington National Press.

JENKIS, C. (1972) Inequality: A Reassessment of the Effect of Family and Schooling in America. New York: Basic Books.

KAHN, H., W. BROWN, and L. MARTEL (1976) The Next 200 Years: A Scenario for America and the World. New York: William Morrow.

LEVEN, C. (1978) "Organizational decline and cutback management." Public Admin. Rev. 38, 4: 316-325.

MOYNIHAN, D. P. (1978) "The politics and economics of regional growth." Public Interest 51 (spring): 3-21.

MELZER, A. H. and S. F. RICHARD (1978) "Why government grows (and grows) in a democracy." Public Interest 52 (summer): 111-118.

MOSHER, C. F. and O. P. POLAND (1964) The Costs of American Governments. New York: Dodd, Mead.

MULLER, T. (1977) "Service costs in the declining city," pp. 119-131 in How Cities Can Grow Old Gracefully. Subcommittee on the City of the Committee on Banking, Finance, and Urban Affairs, House of Representatives, 95th Cong., 1st Sess. Washington, DC: U.S. Government Printing Office.

NACHMIAS, D. (1979) Public Policy Evaluations. New York: St. Martin's.

NORRIS, D. T. and K. HESS (1974) Neighborhood Power and New Localism. Boston: Beacon.

OATES, W. F., P. H. HOWRERY, and W. T. BAUMAL (1971) "The analysis of public policy in dynamic urban models." J. of Pol. Economy 79 (January-February): 142-153.

PETERSON, G. E. (1978) "Fiscally distressed cities: what is happening to them," pp. 83-90 in Local Distress, State Surpluses, Proposition 13: Prelude to Fiscal Crisis on New Opportunity. Hearings before the Committee on Banking, Finance, and Urban Affairs, 95th Cong., House of Representatives. Washington, DC: U.S. Government Printing Office.

ROBERTS, P. C. (1978) "The breakdown of the Keysian model." Public Interest 52 (summer): 20-33.

SUNDQUIST, J. L. (1969) Making Federalism Work. Washington, DC: Brookings Institution.

SCHUMACHER, E. F. (1973) Small is Beautiful. New York: Harper and Row.

WAGNER, A. (1883) Finanzwissenschaft (3rd ed.). Leipzig, Germany: C. F. Winter; also see excerpt from R. A. Wagner in R. A. Musgrave and A. T. Peacocks [eds.] (1958) Classics in the Theory of Public Finance. London: Macmillan.

2

The State of Cities in the Second Republic

THEODORE J. LOWI

☐ CALIFORNIA'S PROPOSITION 13 may prove to have been the shot heard round the world. The movement was bona fide in California, and the Minute-men are rallying throughout the United States. We are witnessing more than a general reaction against the growth of government, and the end is not yet in sight. It will not be my purpose in this paper to discuss taxation, tax limitations, or referenda; rather, the primary objective of this essay is to put these issues in context, so that we might all be more productive of some new ideas.

The California referendum was only one indication that the day of the city will soon be over. The day of local autonomy and local self-government is gone, or soon will be. Since the sun had been setting for a while, we should not have been surprised by the success of Proposition 13 even though it was an ugly and irresponsible way to recognize and hasten the decline of local government. Initiatives like Proposition 13 reinforce Madison's view in the *Federalist Papers* that republican government—i.e., indirect democracy—is superior to direct democracy. But the fact of the matter nevertheless is this: Jarvis and Gann were only accelerating an existing trend. They did not produce the trend.

Proposition 13 is only one of a series of adjustments we are making to a new political system in the United States—a historic break with the past. I call the new system the Second Republic, in order to emphasize the degree of discontinuity. And it is recent. It is a product of changes begun earlier in this century but only culminating in the past two decades. Since the national "constitution" has changed, a change in the position of cities was inevitable. From the standpoint of the cities, the Second Republic represents a change from a federal to a unitary system. Let me trace this out.

In the traditional republic, that is, during our first 150 years, neither the federal government nor the local governments did much governing. Most of the fundamental social choices in our society were made by the states. Even a superficial review of the consolidated laws of any of the American states in 1840 or 1880 or even 1920 will show that the states were the linchpin of the American federal system. The states made all the property laws, the family and estate laws, the public health and safety laws, the labor laws, the laws governing occupations and professions, the credit and exchange laws, the banking and insurance laws, the laws of corporations including municipal corporations, almost all the criminal laws, and most of the other fundamental laws that provide the legal basis of modern life. This includes capitalism itself, which has no legal existence whatsoever except in the state laws of property, corporations, exchange, contract, trespass, fraud, and the like.

A glance at the annual session laws of the U.S. Congress in those same sample years will show how very little the national government had to do. Examples of federal domestic policy include land grants to settlers, land grants to railroad companies, tariffs to certain protected industries, subsidies to the merchant marine, and a sprinkling of regulations around the international ports. Most of the this stuff amounted to a federal policy of husbanding resources and facilitating commerce.

The policies of cities in the traditional republic were surprisingly like those of the federal government. Cities were then and are now constitutionally the creatures of state government and were created and exist for the purpose of carrying out state laws. This status turns out to have had particular meaning in the real governmental life of cities of all sizes. As conveniences of the state governments, local governments were spared the responsibility of making their own fundamental social choices. These fundamental social choices were made by the state legislatures and courts, and many could then be applied according to local variations by the local governments. Generally, this came down to two basic local government functions: The first of those was to build and maintain a variety of essential public works, such as streets and bridges. The second was to provide exceptions or "variances" to the generalities of state laws. Thus saved from the agonizing burden of making the social choices for themselves, cities could treat the resources and the restrictions within state law as privileges that could be handed out on an individual basis as a reward for services or political support. Because of similar policy outputs the local governments and the national government developed strikingly similar political patterns: Any coalition powerful or popular enough to get itself elected could then use the resources of the government to maintain its coalition (Lowi, 1963: Chs. 7, 9; Beer, 1973). This explains away the so-called paradox that national politics remained immensely stable in face of a highly dynamic and changing society and economy.

This same political stability in the midst of social instability has also been a characteristic of American cities. For example, political machines abounded. And machines are nothing more than highly stable political parties. Conditions in American cities were conducive to highly stable political parties precisely because the only policies local political parties sought from government were these subsidy or patronage policies that required no fundamental choices and therefore no broad or integrated social appeals to a public. An amazing number of cities developed advanced and stable political parties, indeed machines. But even those cities west of the Mississippi, where stable political parties did not develop, nevertheless experienced relatively stable local politics. It was simply based upon corporations and trade associations and their coalitions. Railroad, mining, and other corporations in California and other areas of the Southwest actually performed essentially party functions. For a while there was even a union-based machine in San Francisco.

This traditional republic was well over one hundred years of age when things began to happen that would ultimately transform it. The federal government and the cities began to take on new functions; and their politics began to change accordingly.

At the local level, home rule may be the best known governmental change. But of far greater significance was—for lack of a better term—"functional home rule", that is to say, the delegation by state legislatures of specific powers to cities *to make their own fundamental policies* in specified areas. Beginning with the delegation of power to local governments to abate nuisances, delegation from states to cities was expanded to include granting to cities the power to pass their own ordinances regulating slaughterhouses, regulating and establishing public transportation facilities, providing for new street openings and trade centers, ordinances providing for tenement house construction and fire-prevention standards, and a variety of other important powers. States continued to set limits on the breadth of some delegations of power to cities, and so did the courts with their continued application of the Dillion Rule (Adrian, 1976: 117-125). But these limits were successively weakened in practice as the states granted additional powers to cities—powers over local public education, powers over public health, broader powers to make their own policies about local markets, broader powers to pass local laws dealing with land uses, building construction, air pollution, and so on. And as cities modernized by taking on more and more of the power to make their own fundamental choices, cities also *modernized their policies.* Although changes varied, just as the traditional forms varied, there were some general tendencies. One of the most significant of these was the crumbling of stable political parties and other coalitions. There was a general movement of gravity away from electoral organizations toward professional bureaucracies. Politics moved away from part-centered and corporate-centered elites to bureaucratic-professional elites (ct., Dahl, 1961; Swazy and Kaufman, 1960; Banfield and Wilson, 1963). The older structures did not necessarily disappear, but their domains were severely reduced as room was made for new bases of power and new forms of politics. In some of my other writing I have tried to summarize this process of urban political modernization as a move from the old machines to the new machines, with bureaucracies in various forms being the new machines (Lowi, 1968).

Despite the improvements of the level of education and skills of city civil servants, this political modernization weakened the cities in at least three ways. First, it seriously reduced the ability of cities to develop a firm and clear consensus for public policies. Every policy must be forged out among leaders with independent and legally protected bases of power who belong to no common political front. In this sense, it is quite possible for cities to be *well run but ungoverned.* The result is either stalemate or elephantiasis—that is to say, no action or disproportionate expansion of public policies wherever some kind of consensus can be forged out.

The second way in which political modernization weakened the cities is that it reduced the domain or reach of public policy. Even if cities are able to make policies for themselves, those policies are unable to reach many of the sources of city problems because those sources lie beyond the boundaries of cities. Delegation of power to the cities has balkanized the metropolis. The parts add up to something far less than the whole.

A third way this modernization weakened the capacity of cities to govern themselves is that delegation of powers to cities reduced the size of the arena within which locally organized interests would have to play. In relative terms, delegation of power to cities actually increased the power of locally organized interests because now they did not have to go so often to the state capital to compete. Local industry, construction, the building trades, and, especially, the city bureaucracies and their unions, suddenly found themselves able to bargain from a position of strength with the mayor. The local private interests loomed larger and larger at the very time the mayors found themselves more and more isolated from their independent, popular base. Mayors throughout the country had to make their peace with local interests, especially their own bureaucracies.

All three of these developments created tremendous imbalances. Demands by organized economic interests were looming ever larger just as these were joined by the demands of hitherto underprivileged groups in the cities. Within this power imbalance most cities found themselves literally too small to handle their policy problems but politically too weak to resist trying. It was inevitable that city governments would respond by pushing their budgets and their tax bases to, and beyond, their limits. City leaders did this because they perceived that they had no choice; they did it because their electoral conditions were predicated on expansionism; they did it because they could say to themselves, "Après moi, le déluge"; and they did it because they had been encouraged since 1946 to believe that ultimately the national government would pick up the tab. Thus, the special fiscal problems of the cities in the 1970s are a new twist to problems that go back a good way.

In the East, it had begun as early as 1946, and the problems accelerated after that. But the makings of the same expansionism, with similar results, are already to be observed in many prosperous cities of the Sun Belt and the West. Political scientists David Perry and Alfred Watkins in a New York *Times* op-ed article last April reported that a higher percentage of ghetto populations in the eleven largest Sun Belt cities live on incomes below the BLS minimum than those who live in ghettos in the largest cities in the supposedly poverty-stricken Northeast. The difference is that for the moment a larger proportion of the inner-city poor in the Sun Belt are holding jobs. If their jobs pay so little that their incomes are less than the minimum average family budget, this indicates only that there is a better source of cheap labor for corporations in the Sun Belt than in the Northeast. But, ultimately, if present trends continue, most of these cities will also be pushed to their limits because (1) they too suffer from the three problems of modernization identified above, and (2) there is no indication that they will be able to draw sufficient surpluses from their highly stratified populations to maintain public services, social services, physical plants, and security of persons and property. Concessions already made to attract new industry are heavy liens against future revenues, and corporations can blackmail new cities as well as old ones with threats to move. Increases in interest rates and debt service payments in many of these cities suggest that the reckoning day was not distant even without Proposition 13.

During pretty much the same period, the federal government also took on new functions and in so doing began to look less and less like the traditional national government we had grown to know and love. Although there were antecedents, the big turning of the corner was the 1930s, when we vastly expanded the tradition of subsidizing commerce and added two entirely new kinds of functions, new at least for the federal government. These functions were *regulation* and *redistribution.* In adopting a large number of regulatory policies, the federal government was exercising powers that supposedly had been reserved by the Constitution primarily to the states, that is, the national government was asserting its own "police power." As for the second of the new functions, conventional labels referred to fiscal and monetary policies, but these bland labels masked the true significance and novelty of the federal government's new redistributive functions. Although regulatory and redistributive policies are quite different from each other, they share one very important attribute that distinguishes them from almost everything the federal government was doing prior to the 1930s: These two new functions involved the federal government in *direct and coercive use of power over citizens.*

The return of the Democrats to power in 1961 was the real signal of the beginning of the Second Republic, because the return of the

Democrats was accompanied by an entirely new attitude. If the 1930s had established a strong national state as politically feasible and constitutionally acceptable, the 1960s made the strong national state a positive virtue, desirable for its own sake. Well before the end of the Democratic era of the 1960s, we had established a full-scale modern national state in the United States. A national presence had been established in all areas of social and economic endeavor. For the first time in the history of the United States, the national presence had extended directly to local units, to individual citizens, and even into the streets.

You can look at virtually every commonplace fact about the institutions of American government and politics, and you will find that during the 1960s and 1970s their relations to each other have changed. The Second Republic is simply my name for that fact, and for the fact that we have looked at all those changes and found them good. We now have a new republic just as much as if we had sat down and held ourselves a new Constitutional Convention.

Among those changing relationships are a number concerning urban life. As we were becoming a new republic, the cities appear to have gone through three changes of status. The first of these, more or less the traditional status, was a true federalism stage. Cities were far apart from the national government, in practice and by constitutional principle. They were nestled within each state and were left free to go about their merry way.

The second phase was what I would call a three-layered federalism, and maybe the marble cake was an appropriate imagery after all. It is during this phase that the federal government became very active in what has been called "urban policy," but these urban policies were structured so that the cities continued to remain quite autonomous. Most urban policies during this period were "categoric grants," where the federal government did set conditions on how the money was used, but within these conditions the cities got the money if they could prove themselves eligible. From then on it was theirs. But during that very period, cities began to develop direct relationships to Washington, and organizations of mayors and other municipal officials became very effective lobbyists without bothering to go through the intermediary of state government. This signals the beginning of the third phase.

The third phase, which takes us into our own era, is too recent to be fully settled, but certain of its features can be seen. I would call this the unitary phase, and I choose that concept with care in order to indicate that this is a move directly away from traditional federalism toward a more direct European-type federal-city relationship. It involves not only the bargaining and lobbying directly between city and Washington, but it now also involves sustained and hierarchical bureaucratic relationships. The cities are now operating increasingly as agents of the federal government. There had been signs of this change toward the unitary system for quite a while. However, the War on Poverty is to me the real beginning of this new relationship. It was a regular fiscal relationship for at least four reasons: it involved direct and everyday relationships between city agencies and federal agencies; these relationships were bureaucratized; the federal government maintained a tremendous amount of discretion over the actual behaviors of city agencies, particularly how their funds were used; and city agencies plus locally based interests grew accustomed to regular relationships with the federal government, including careful surveillance. This third or unitary phase signals the beginning of the Second Republic for the cities. It will be found in the expansion and bureaucratization of federal discretionary social policy.

Direct and bureaucratized federal-city relationships spread from the War on Poverty to the various other categoric problems of urban policy. As these became increasingly discretionary, so did the relationships between cities and the federal government become more and more intimate and regular. President Nixon's fundamental program of revenue sharing and his secondary programs of "specialized revenue sharing" extended and more firmly established this new, unitary relationship between cities and the federal government. Many will argue that revenue sharing was Nixon's way of appearing to be liberal while at the same time cutting back on federal support of cities (Nathan et al., 1977). This may have been true, but there is something far more significant about the fact that revenue sharing took the categoric urban programs of the Democrats, which were already highly discretionary, and made them completely discretionary. This meant that the president or the Treasury Department could retain a regular and everyday fiscal power over the cities. This also meant that the federal government could use federal money on a patronage basis to tie cities all that much closer to the national government. The fact that the total amount of federal urban money was going to be reduced only intensified the scramble and tied the cities all that much closer to the national center.

Special revenue sharing accomplished almost the same thing, because these programs consolidated so many previous categories and were so broad and vaguely defined that national discretion over cities can be almost as great as with general revenue sharing.

The final and most significant step in this third phase is the emergence of discretionary fiscal policy in the form of federal investment insurance. Most everyone knows about the so-called Lockheed loan, but few know it was not a loan at all, but merely an official signature on a document with which Lockheed could go to private banks to borrow up to 250 million federally insured dollars at lower rates of interest or not otherwise forthcoming at all. This method of government intervention became a little better known in 1977-1978 because of Congress's decision to underwrite in a similar manner nearly two billion dollars of New York City revenue bonds, which, without that federal government guarantee, would have gone very little distance on the open bond market. This is the federal government's first entry into underwriting the solvency of American cities, and it is almost certainly not going to be the last.

Investment guarantees have been in use for some time, but grew so spectacularly in the 1960s that this has become the most important discretionary fiscal weapon in the hands of federal bureaucrats. Moreover, the important examples of the use of discretionary investment guarantee authority are from programs of particular relevance to cities. These include the Veterans Administration housing programs and the housing programs under the FHA, urban redevelopment and community facilities, small business administration, including the black capitalism encouragements, and low-rent housing. The value of new private investment that received federal guarantees in 1976 came to $53,463,000,000. The total value of all outstanding private investments that were enjoying federal guarantees in 1956 came to $243,213,000,000. A conservative estimate of the 1978 cumulative value of private investments enjoying government investment guarantees was $323.5 billion (Council of Economic Advisors). And bear in mind that this figure represents only discretionary investment guarantees—therefore not including such other government investment guarantees as the Federal Insurance Corporation and the Federal Savings and Loan Insurance Corporation. These figures also do not include investment guarantees issued by such semipublic intermediate credit organization as the Federal Home Loan Insurance Corporation (Fannie

May) and the Government National Mortgage Association (Ginnie May). To give you some sense of the present scale of this kind of government activity, the total value of all investments enjoying government guarantees annually since 1976 has almost exactly equaled the total gross private domestic investment for each of those years. For example, total new investment in 1976 was $243 billion, and the total outstanding value of discretionary investment guarantees was $243.2 billion (Lowi, 1979: Ch. 10). The private investment figure is of course the actual investment for that year, whereas the government guarantee figure is the cumulated outstanding investment guarantees. However, the closeness of the two figures indicates that the sector of investment guarantees has come to be a kind of balance wheel in the economy. Government underwriting of risk is as important as it is unappreciated in our economy.

As cities tie themselves closer and closer directly to the federal government, certain consequences tend to flow. The first of these is the muddying of lines of responsibility and accountability. Cities and their officials organize as lobbyists in Washington. City bureaucracies tie themselves to counterpart bureaucracies in Washington. Surveillance follows. Now a certain degree of lost local autonomy might be welcome. But federal surveillance of these highly discretionary programs is bad surveillance. The Treasury cannot even do a metriculous job on New York. Multiply that by ten or a hundred. Lines of authority become disjointed. The distance between taxing units and spending units is stretched still further. And this is compounded by the fact that federal investment guarantees are totally "off-budget," never reaching Congress or the Office of Management and Budget (OMB). We may lose entirely our ability to fix responsibility.

A second highly probable consequence of federalizing the city is the expansion of private power over cities—especially the power of the finance community. Already at a maximum inside the cities—as described above—federal bond guarantees in particular will throw more of the power over cities to corporations. Mayors may be especially susceptible to deals that promise to displace further local financial burdens.

A third consequence is a highly probably change in the distribution of city services. Since cities are confronted with declining revenues and also with declining direct assistance from all outside sources, they will have to cut most heavily into those local policies that are financed by current taxes or subsidies. And the local services and activities most likely to be maintained are those that can be maintained through borrowing by the sale of local revenue bonds, eventually by sale of local revenue bonds that are guaranteed like New York City's bonds by the federal government. This means that the long-run bias of city activities and services will be toward public works and pension maintenance. This will be true if the present structure of welfare and social security remains, and it will be all the more true if the federal government finally picks up all of the social security and welfare system. The implications of this are as follows, as far as I can tell: public works are largely for the business and the stably employed community. Pensions are for the entitled bureaucracies, the organized city bureaucracies.

Are we prepared to accept without a fight the drift of cities in the Second Republic? It seems to me it is at least worth trying to head it off because city governments are combining the worst of the modern city with the worst of the traditional city—that is to say, responsibility for local regulatory programs plus the dispensation of public works and public patronage, yet without the full legal or financial responsibility for either. If our cities are traveling the route toward a unitary system within the Second Republic, I would propose drastic intervention rather than the marginal salvaging operations called "retrenchments" in which we are preparing to engage ourselves today. And referenda are simply putting us deeper in the trenches, contributing still further to the situation in which the cities have one foot in the traditional and one foot in the new, and the worst of both.

The intervention I would propose is a fundamental, constitutional redefinition of the city. I would take back all the charters and write new ones with far fewer powers and responsibilities, leaving cities only with those few functions that can be shown to be inherently local. Don't even try to salvage the cities as we know them. Metropolitanization has already made traditional local government obsolete, just as geographic nationalization of the economy made traditional—i.e., limited—national government obsolete by the beginning of the twentieth century.

Frank Lloyd Wright was once asked what he would do about New York City, and by implication all cities. Without hesitation, he replied that he would start all over again—at the second floor. Maybe we should take a page from Wright's book and start all over again at the second level. I propose, therefore, that we explore that neglected and unappreciated unit of regional government, the American states. As the Queen told Alice, believing the impossible only takes practice. Lately I've been practicing, and I can report to you that it is getting easier and easier to believe that state government may be a solution. Retrenchment is certainly not the answer. Get out of the trenches altogether. Retreat, but to higher ground.

REFERENCES

ADRIAN, C. (1976) State and Local Governments (4th ed.). New York: McGraw-Hill.
BANFIELD, E. and J. WILSON (1963) City Politics. Cambridge, MA: Harvard University Press and MIT Press.
BEER, S. (1973) Publius: 49-95.
DAHL, R. A. (1961) Who Governs? Democracy and Power in an American City. New Haven, CT: Yale University Press.
HUNTINGTON, S. P. (1968) Political Order in Changing Societies. New Haven, CT: Yale University Press.
LOWI, T. J. (1969) The End of Liberalism: Ideology, Policy, and the Crisis of Public Authority (2nd ed.). New York: W. W. Norton.
――― (1968) "Foreword," in H. H. Gosnell, Machine Politics: Chicago Model. Chicago: University of Chicago Press.
NATHAN, R. P., C. F. ADAME, Jr., and Associates A. JUNEAU and J. W. FOSSETT (1977) Revenue Sharing: The Second Round. Washington, DC: Brookings Institution.
KAUFMAN, H. and W. SWAZY (1960) Governing New York City: Politics in the Metropolis. Washington, DC: Brookings Institution.

3

The Possibilities for Urban Revitalization: Constraints and Opportunities in an Era of Retrenchment

LOUIS H. MASOTTI

PRESENT POLICIES AND THEIR LIMITATIONS

☐ DESPITE SOME RHETORIC to the contrary, urban policy in America, by and large, has been designed to be therapeutic: to treat the symptoms of the urban malaise rather than to uncover its fundamental causes and to engage in preventive action. Since the rediscovery of the urban crisis in the early 1960s, our approach has been to develop more programs, create more bureaucracy, and spend more money in an effort to stem the tide of urban decay. For the most part, this behavior was a political response to pressure from the cities and their advocates.

The fundamental result of all this activity has been the development of programmatic, bureaucratic, institutionalized policy failures. In general, the cities and their residents are less well off now than they were before the flurry of urban policies since 1949. Those policies have been ineffective and inefficient in pursuit of the ostensible goals—the improvement of the quality of urban lives, and the support of cities as economic entities. Our urban policies, which currently use terminology such as "targeting and tailoring" for "distressed cities," operate primarily as a maintenance function. They are designed to prevent things from getting worse rather than to help things get better. Urban policies

AUTHOR'S NOTE: *This is a revision of a paper entitled "Toward a viable urban future in a society of limits: possibilities, policies and politics," delivered at the 1978 American Political Science Association meetings in New York City, August 31, 1978. The author wishes to acknowledge his intellectual debt to John L. McKnight and Joy Charlton and to express his appreciation for the editorial work done by Vivian Walker and Georgette Carlson.*

tend to be dependency-producing—for urban populations in relation to their governments, and for cities vis-à-vis the federal agencies. Such dependency significantly reduces the possibilities for the emergence of either effective, self-sustaining citizens or for viable cities. This condition, it has become clear, is acceptable to neither the recipients of urban assistance nor to its critics, but for different reasons. The recipients are dismayed that extant policies are not effective in terms of their purposes, while the critics tend to emphasize programmatic inefficiency. Both are correct.

In short, there seems to be ample evidence that current treatment-maintenance-dependency policies and programs are not working. Let me cite some evidence from three critical urban policy areas—health, education, and security. We have dramatically increased our commitment to medicine without significant improvement in the quality of health (measured by morbidity and mortality rates); we have increased our budgets for police (professionalization, new equipment, and technology)) in return for rampant crime and a growing sense of insecurity; and in education, we have focused resources on school facilities and elegant curricula only to discover dramatic decreases in the quality of education (measured by scores on standardized tests). In addition, it is becoming apparent that other institutions in our urban society charged with responsibility for a variety of needed functions are failing to perform them effectively—the prisons, the mental health institutions, the child welfare system, and gerontological programs.

More importantly, however, as we look toward the future it does not seem likely that an approach to the urban condition which has operated in an era of expansion focused on *increases* in budgets, personnel, and institutions can function effectively as we become increasingly a "society of limits." The environmentalists and the Arab nations have demonstrated limits on our natural resources, and California has dramatically led the way toward the probability of significant decreases in public financial resources available in the future for urban, as well as other, problems. It is imperative for us also to understand the current limits of our institutional imagination and thus the possibilities for developing effective policies, and the even smaller probabilities of implementing them. Institutions with an urban orientation—both public and private, national and local—engage in self-perpetuation and self-maintenance despite the emergence of such accountability-encouraging techniques as "sunshine" and "sunset" laws and zero-based budgeting. Institutional bureaucrats have been able to circumvent such efforts at accountability by assigning to themselves the roles of need-definition and client-identification. In addition, institutional bureaucrats have been effectively protected either by civil service programs which sanction imagination more often than incompetence, and by public service unions which give the impression that they exist to stifle the productivity of their members.

Thus, the very institutions, professionals, and bureaucrats which under present urban policy are charged with alleviating or mitigating urban problems, or at a minimum, with maintaining the status quo, have become counterproductive, inefficient, ineffective, and very, very expensive; in short, they have become more disabling than enabling.

At the same time, public awareness of such inefficiency and ineffectiveness has led to increased alienation from both governments and their programs, to a reduced level of tolerance for incompetence and inefficiency, and to increased demands for more accountability, productivity, and a higher quality of service. Encouraged by continued rampant inflation, which reduces their purchasing power and the quality of their lives, the middle class has joined the more disadvantaged in protesting the shortcomings and the failures of governments. The one they can do most about—local governnment—is feeling the brunt of this disaffection, although its effects are now being felt in the state house, in the White House, and on Capitol Hill. The basic demand is for more service and fewer taxes or for fewer services and reduced taxation, both of which clearly set the stage for major political conflicts and the potential for significant political realignments. We are used to getting less for more in this nation; it will be more than a little interesting to see how we handle demands that there be "more for less," or even "less for less."

RESETTING THE CIVIC AGENDA

In order to explore the possibility of a viable urban future in America, it becomes necessary to clarify or redefine the goals of an acceptable or desirable urban condition. In general, this involves improving the quality of lives, developing appropriate procedural arrangements for effective citizen participation, and fostering a meaningful sense of community. More specifically, I would contend that a viable urban future requires policies which promote vital communities of healthy, educated, employed, and decently housed citizens secure in person and property, with reasonable opportunities to participate effectively in the governing process, and with reasonable expectations that both public and private sector policy-makers and bureaucrats will function under fundamental rules of equity and justice. The possibility of achieving such goals, given our failures to date and the impending imposition of new limits on resources, is a considerable challenge to our conceptual ingenuity and our sense of commitment. The problems we confront as we go about resetting the civic agenda include reconceptualization of the urban problem, imaginative use of increasingly limited resources, appropriate procedures to ensure effectiveness, and ongoing evaluation of both intended and unintended consequences.

Clearly, more of what we have done in the past is an inadequate way to approach the future. It becomes necessary to explore options and alternatives in policy and procedure for the potential achievement of the stated goals. This process includes at a minimum:

(1) an evaluation of the appropriate allocation of functions in contemporary urban society, and
(2) the functional "capture" of current trends, movements, experiments, moods, and themes which might be put to positive use in an effort to enable and empower individuals, groups, neighborhoods, and cities to increase the probabilities of a viable urban future in an era of resource limitation.

The necessity of asking some tough questions and finding appropriate answers to them is obvious. It is imperative that we begin to think about policies that are both different and more effective. If we are agreed, at least in general, that it is the quality of lives that individuals lead in their local communities that is or should be our primary concern, and that we have been unable or unwilling to organize ourselves effectively, to design imaginative policies, and utilize *all appropriate resources creatively* in an effort to move toward a condition of citizen and community well-being, our choices are three:

(1) to *capitulate,* leaving the cities to atrophy and to become "sandboxes" or "reservations" for the disadvantaged and have-nots, which is clearly unacceptable;
(2) to continue our *present policies and procedures* which have led to dependency and have proven so ineffective in a period of growth as to project increased levels of failure in a society of limits; or

(3) to express our dissatisfaction with the current urban condition and, rather than wringing our hands in despair, to set about the task of *reconceptualization, reorganization,* and *recommitment* to the most effective use of the vast personal, community, and governmental resources available to us.

The real contemporary challenge to the cities, and especially to their leaders, both public and private, and indeed to their citizens, is to design imaginative policies which use scarce resources more creatively and thus energetically to pursue policies which do as much or more with less.

This pursuit requires significant adjustments and alterations in the definition of our urban problems, in the allocation of authority, responsibility, and accountability, and in the evaluation of policy success and failure. The federalization and centralization of government and the concentration of corporate economic power has resulted more often than not in the stifling of imagination, the rigidity of policy, and the growth of institutional dependency and counterproductive bureaucratization. Increasingly the measures of success have become budget bottom lines and/or the size and the complexity of the institutions to which they are assigned, rather than the consequences and impacts which dollars, programs, technology, and organization have on the problem addressed. We have become enamored with programs more than productivity, and with institutions more than impact. Our policies too often are structure- rather than solution-oriented. Our priorities have been misplaced and must be replaced with those more attuned to the real needs of real people in real communities.

We must be about the business of decentralizing, deinstitutionalizing, and deregulating where and when it can be demonstrated that such actions will accrue to the benefit of citizens, clients, students, patients, and consumers in particular, as well as to cities in general. Schools which do not or cannot educate, medical facilities which are iatrogenic rather than health-giving, welfare systems which blame the victims rather than enhance their opportunities, and police departments which are unable to secure people and their property must be reorganized in ways which will improve the quality of individual and community life.

In a society of limits we must begin the process of "enabling" and "empowering" individuals, neighborhoods, communities, and organizations to help themselves by facilitating the flow of private capital and public funds and professional/technical assistance where such resources will improve the quality of life through the active participation of those affected. There is some evidence to indicate that such programs already in force are not only more policy-effective but also more efficient in the utilization of scarce resources, particularly where labor intensity and preventive behavior displace fiscal intensity and treatment. In short, decreasing institutional/bureaucratic dependency and encouraging the building of capacity at the level of the local community might well move us in the direction of both program effectiveness and cost efficiency at precisely the time when each is being defined as a significant issue. The key question is not *who* makes social policy or *where* it is made but what difference the policy makes in the lives of those it is intended to help and on the life of the community itself. By questioning current assumptions about the determinants of the quality of life in this society it becomes possible to identify those interventions—public and private, institutional and individual, short- and long-range, small- and large-scale, professional and voluntary—which are judged most likely to make a difference, to pursue those policies purposefully, and to evaluate their effects. In this process it is important to identify the most appropriate role for governmental institutions by level and policy area, and for the private sector whether it be corporation, financial institution, service organization, neighborhood, family, or the individual.

The potential success of such an approach to urban policy in an era of limits will be based on our capacity of identify *real problems,* rather than bureaucratic and professional definitions of problems, and to design intervention strategies which employ the most effective and imaginative combination of public and private resources in terms of the needs of those to be served and of the community of which they are a part. In order to do this, it will be necessary to take advantage of existing and emerging tendencies designed to overcome dependency and to make most effective use of existing resources, opportunities, and procedures.

Ivan Illich, for example, has argued convincingly against counterproductive technology, for the deinstitutionalization of education and criticized the medical industry as a "nemesis." The disabling and repressive effects of professionalism have been argued persuasively by Eliott Freidson and John McKnight. In the serviced society which we have become (more than two-thirds of us now derive our income by delivering services), there is a dilemma in the growth of the service economy. In order to serve one another we need more clients who need help or clients who need more help. In a society served by professionals and bureaucrats, people are increasingly defined as lacking, disabled, or deficient. There is a need for need and professionals are entrusted with the responsibility for defining that need which is, in turn, self-serving, dependency-producing, and expensive. A growing serviced society depends on more people who can be defined as problems rather than as productive participants. A professionally dominated society/economy thrives on the increase in the deficiencies of the population rather than on its potential to develop a capacity for well-being. Even the consumer/voter revolt, symbolized by Proposition 13 and its offsprings, does not address itself directly to the issue of professional dominance and client dependency. It is more an expression of alienation and disaffection for the cost-effectiveness of products, services, and programs. In effect we are hearing the protests of the "middle-class poor" who want reliable products and accountable public officials.

There are, however, some indications of capacity building, enablement, and empowerment which portend the possibility of viable cities and populations. Berger and Neuhaus offer moral, political, and practical reasons for supporting and enhancing "mediating structures"—neighborhoods, family, church, and voluntary organizations—as effective bridges between individuals and societal megastructures. They argue forcefully that such mediating structures be protected and fostered by public policy as essential ingredients for capacity building and, therefore, for effective democratic procedures. But they also submit that such structures can be utilized effectively by public policy-makers to realize social goals. The mediating structures are, in fact, an argument for dissembling large scale public and private power and resources for more effective and efficient use.

Others, including Schumacher, have made a case for redesigning tools and technology for more effective utilization at the community, familial, and individual level. The so-called "appropriate technology" approach insists that economies of scale are not always economic and very often counterproductive. The Center for Neighborhood Technology (CNT) was recently established in Chicago to act as a clearinghouse for new ideas for utilizing technology to enhance family, neighborhood, and community well-being. CNT makes a concerted effort to identify the diversity of technological innovations available, such as cost-efficient household solar energy and roof top hydroponic greenhouses to improve the supply of inexpensive fresh vegetables in a lower-class black neighborhood. Other "appropriate technologists" in the Chicago area have argued forcefully that the $8 billion Deep Tunnel Project of the Metropolitan Sanitary District (the largest public works project in the history of the United States) is not only outrageously expensive but, when finished, is not likely to reduce the flooding problems for which it is designed; they argue instead for the development of water retention devices located locally with a provision for recycling on site.

Appropriate technology is only one of several facets of the new interest which has emerged focusing on neighborhood viability. The National Commission on Neighborhoods is at least of symbolic significance in this regard, although the dramatic increases in neighborhood revitalization throughout the country are much more important. Neighborhoods are being upgraded in many cities because of newly available mortagage money resulting from successful antiredlining legislation, and by the rehabilitation and redevelopment which is brought on by "regentrification" (especially in cities like Baltimore, Washington, and San Francisco). Neighborhood revitalization is a good example of the combined effects of resident politicization and effective regulation by appropriate governmental agencies. Since people live in neighborhoods and not in cities, the renaissance of those neighborhoods bodes well for urban viability.

Another area of low-cost/high-effectiveness activity which reduces dependency on professionals and bureaucrats is the so-called "self-help" phenomenon which has grown simultaneously with the disaffection with human service bureaucracies. By banding together in groups whose members share similar problems, i.e., through mutual assistance, the self-help groups have achieved considerable success across a wide variety of purposes—from the venerable Alcoholics Anonymous to organizations for cancer patients, and an endless variety of physical, mental, and emotionally distressed persons. Current estimates indicate that there are more than one-half million self-help groups in the United States comprising more than fifteen million members. These groups are characteristically developed outside of established institutions by persons sharing a characteristic or a problem who have agreed to take responsibilities for themselves and to help others do the same. Self-help is for the most part a nonprofessional, nonbureaucratic, grass roots response to the increasing inefficiency of our complex, technological society.

Renewed interest in mediating structures, appropriate technology, neighborhoodism, and self-help groups results in part, at least, because of dissatisfaction with the options. Such activity is an expression of antiinstitutionalism, antibureaucracy, and antiprofessionalism and constitutes a sort of "service populism" which has significant impact at the personal and small-scale organizational level. The critical question may be whether these small-scale, cost-efficient, high-efficacy efforts at providing advice, aid, and service to millions of people can be and should be a substitute for, or a complement to, existing large-scale bureaucratic, social, and human service programs in both the public and private sectors. Such activities seem to fill the vacuum created by the limits and failures of established bureaucracies and that may, in fact, be their strength. It does seem somehow inappropriate to try to transform highly successful small-scale programs into organized service delivery programs; perhaps they work best because they operate on the basis of felt need and outside the established bureaucracy. It is tempting to suggest that, since such programs seem to have real potential for effectiveness, they be substituted for existing, inefficient programs in the public sector. But because it is clear that such programs work because they are not bureaucratic, institutionalizing them would merely undermine them rather than transform them into effective large-scale policies and programs.

What is suggested and possible, however, is a careful reconsideration of how private interests and purposes can be "leveraged" in the public interest. Governments will have to begin to think more seriously about using their authority to regulate and to use tax expenditures (subsidies, loans, credits) rather than spending reduced public dollars on maintenance and income services and on incentive-reducing capital grants. Governments must increasingly use their authority to provide incentives for the private sector, including groups and individuals, to do things in their own best interest, which are also in the public interest. In Chicago recently, the city was able to loosen the mortgage money market considerably and to provide an incentive for home buying for middle-class families by authorizing $250 million in municipal bonds to be used by financial institutions as subsidies in reducing home mortgage rates from 10 to 8%. Chicago's financial institutions were more than happy to comply; it cost the city nothing; and it has created a significant increase in housing investment in the city at precisely the time when high mortgage interest rates were stifling the market.

THE IMPLICATIONS OF URBAN ALTERNATIVES: POLITICAL, SOCIAL, AND ECONOMIC

All policies have costs and benefits. We are concerned here with those associated with thinking small, creatively, and effectively. The implications of shifting urban policy in the direction of the private sector include at least the following: the possibilities for new political coalitions, the impact on political forms and processes, the practicality of disaggregating responsibility for the quality of urban lives, and the problem of maintaining our commitment to the disadvantaged.

One of the most interesting possibilities suggested by the urban policy options suggested above include the potential for a liberal/conservative coalition. Liberals interested in more effective urban policy, neo-conservatives concerned with more efficient policy, and classical conservatives who want ideological purity might find some agreement if effective policies can be identified which both reduce the role of government and the expenditure of public dollars. Whether such coalitions come to pass depends on how well the policies are articulated and on the nature of their nuances.

The policy options we have been discussing involve a decentralization of responsibility and control. This may involve the decline of order and accountability in the world of public policy. Public officials, unsure of their role in a public/private partnership, may resort increasingly to buck-passing techniques such as the referendum. There are real questions associated with a shift from indirect citizen participation through elected officials to the kind of town meeting democracy which Proposition 13 and the 1978 Cleveland mayoral recall election portend. Complex social legislation may be more difficult than the populists would have us believe since direct democracy removes the possibility of compromise and bargaining during the policy process and places the responsibility for deciding the problem of interpretation and implementation in the hands of the very bureaucrats at whom referenda and initiatives are directed. One of the more interesting side shows following the passage of Proposition 13 is the kind of "we'll show you" revenge evidenced among some California public agencies which were potentially damaged by the loss of tax revenue. Under the circumstances, elected officials may be forced to demonstrate their capacity to lead and act responsibly on behalf of the public interest.

There are also some real doubts about the potential for a disaggregated responsibility for social policy in the cities. The rich have always been able to buy services they need and do not necessarily rely on public providers. The poor for the most part have been unable to provide for their own services; they tend to lack adequate knowledge of the system and adequate technological skills. The middle class, which has indicated the strongest interest in self-help and reduced public service as a tradeoff for lower taxes, may very soon tire of "doing it themselves" as the romance wears off and the drudgery sets in. For example, how many of us want to, can, or will sweep our own streets, dispose of our own garbage, supplement a reduced school curriculum, or participate in a neighborhood security watch. The substitution of labor intensity for fiscal expense quickly becomes inconvenient for a society spoiled by service and socialized to expect it. Acceptance of a process of *lowered* expectation rather than rising ones, will not come easily, if at all. The heralded Chinese model of cooperative self-help is a function of intensive socialization, peer group pressure, and government sanction; it is an act of political, economic, and social will, which I suspect the American urban middle class is not yet ready for.

Perhaps the most serious implication of social service disaggregation and self-help concerns the dynamics of maintaining our commitment to the disadvantaged. Most of the social spending or maintenance policies of the government have been directed at the poor, and in many cases it is precisely those programs which are the target of tax resistance and revolt. Initiatives such as Proposition 13 rely on majority disaffection rather than on minority need. Issues of equity, compensatory assistance, and need are not taken into account in most populist techniques. In a short time we have moved from a concern for service equity as articulated in the Rodriquez and Serrano cases to a service equivalent of the Bakke case. Policy options which give vent simply to the felt need for reduced taxation rather than to the more effective delivery of service to all urban residents in need may be judged not only a failure but a disaster. The middle class may need and get tax reductions or stabilization through its political clout, but there must be some assurances that the poor and disadvantaged will be adequately protected and serviced and, if possible, enabled to develop the capacity to do things for themselves in the same way that the middle class has been able to.

CONCLUSIONS: RISKS AND PAYOFFS

If indeed we have approached an era of limits and if there is some potential for viability despite those limits, it would appear that we have three options:

(1) we can do fewer of those things which we have done;
(2) we can do less of the same things; or
(3) we can do the same things differently including the transfer or sharing of responsibility for all or some of them.

It does not seem to me that there are any options to the options. Which of these options are embraced or which combination of them is devised is indeed a political question with serious implications for the future viability of cities, neighborhoods, and populations. An effective set of imaginative urban policies which decreases dependency, increases a sense of community, and develops the potential of citizens and their communities would be welcome indeed. However, the path toward urban viability involving the use of human energy and private capital is not clearly charted. I have long thought that the impending era of limits was a cloud with a silver lining, one which would cause us—indeed, force us—to reconceptualize our problems and solutions in some more meaningful and effective way. I would like to hope that I have not been too optimistic.

REFERENCES

BEAL, F. (1978) "Helping cities help themselves: American urban problems and prospects." Presented at "The Conservative Approach to Inner City Problems," a conference for those concerned with the problems facing our inner city areas, Birmingham, England.

BERGER, P.L. and R. J. NEUHAUS (1977) To Empower People: The Role of Mediating Structures in Public Policy. Washington, DC: American Enterprise Institute for Public Policy Research.

FREIDSON, E. (1970) Professional Dominance: The Social Structure of Medical Care. Chicago: Aldine.

GORDON, A., M. BUSH, J. McKNIGHT, L. GELBERD, T. DEWAR, K. FAGAN, and A. McCAREINS (1974) Beyond Need: Toward a Serviced Society. Evanston, IL: Center for Urban Affairs, Northwestern University.

GORDON, A., M. T. GORDON, and R. LeBAILLY (1976) "Beyond need: examples in dependency and neglect." Presented at the Annual Meetings of the American Sociological Association, New York, New York.

HAIDER, D. (1978) "Urban policy options: where we have been, what we have learned, and where can we go?" Background paper prepared for a Conference

on Viable Urban Futures, April 1978, Center for Urban Affairs, Northwestern University.
——— (1978) "Viable urban futures: a proposal for a national conference and publications." Evanston, IL: Center for Urban Affairs, Northwestern University.
ILLICH, I. (1975) Medical Nemesis: The Expropriation of Health. London: Calder and Boyars.
——— (1973) Tools for Conviviality. New York: Harper and Row.
———, I. K. ZOLA, J. McKNIGHT, J. CAPLAN, and H. SHAIKEN (1977) Disabling Professions. London: Marion Boyars.
KHARASCH, R. N. (1973) The Institutional Imperative: How to Understand the United States Government and Other Bulky Objects. New York: Charterhouse.
LONG, N. (1971) "The city as reservation." Public Interest 25 (fall): 22-38.
MASOTTI, L. H. (1977) "Social policy in America: the need for reassessment." Evanston, IL: Center for Urban Affairs, Northwestern University.
McKNIGHT, J. (1976) "Professionalized service and disabling help." Presented at the First Annual Symposium on Bioethics of the Clinical Research Institute of Montreal.
MOLOTCH, H. (1979) "Capital and neighborhood in the U.S.A.: some conceptual links." Urban Affairs Q. 14 (March).
MORRIS, R. S. (1978) Bum Rap on America's Cities: The Real Causes of Urban Decay. Englewood Cliffs, NJ: Prentice-Hall.
"New partnership to conserve America's communities." (President Carter's urban policy statement.) The White House, March 27, 1978.
Office of Planning and Research, State of California (1978) "An urban strategy for California." Sacramento: Office of Planning and Research.
OLSON, M. (1965) The Logic of Collective Action: Public Goods and the Theory of Groups. Cambridge, MA: Harvard University Press.
——— and H. H. LANDSBERG [eds.] (1973) The No-Growth Society. New York: W. W. Norton.
SCHULTZE, C. L. (1977) The Public Use of Private Interest. Washington, DC: Brookings Institution.
SCHUMACHER, E. F. (1973) Small Is Beautiful: Economics as if People Mattered. New York: Harper and Row.
STERNLIEB, G. (1971) "The city as sandbox." Public Interest 25 (fall): 14-21.
——— and J. W. Hughes (1978) Revitalizing the Northeast: Prelude to an Agenda. New Brunswick, NJ: Center for Urban Policy Research.
"Toward a more effective approach to urban problems: definition, analysis, intervention, and evaluation." (1973) Evanston, IL: Center for Urban Affairs, Northwestern University.
WALSH, A. H. (1978) The Public's Business: The Politics and Practices of Government Corporations. Cambridge, MA: MIT Press.

4

Bureaucratization of the Emerging City

SCOTT GREER

☐ THE CITY THAT HAS EMERGED in the wealthy nations during the last fifty years is something new in history. It has no walls or other container, it sprawls over vast areas, and it fades imperceptibly into suburbs, exurbs, small towns and finally the countryside (Greer, 1962a, 1962b, 1971). It is a disturbing phenomenon, for it has no governmental unity—no "political container." Instead, it is a central core of declining numerical importance and a surrounding sea of suburban municipalities. The city of Newark, for example, has now some 18% of the metropolitan Newark population.

Due to increasing scale of energy conversion and of organizational magnitudes, the city is a meeting place for the headquarters or branch offices of gigantic organizations, public and private. In the dominance of exclusive membership groups—corporations, labor unions, governmental agencies, and the like—the local area everywhere loses the ability to control its own destiny (Greer, 1955). The spatially inclusive group, the city, gives way to giant formal organizations and critical decisions for the city may be made far away by persons who assign little weight to its future.

That spatially inclusive group, which once encouraged integration among exclusive membership groups and small primary groups in something which could be called "the polity" (Long, 1961), is today merely a turf on which contending groups relate. Integrating power is generally provided by giant formal organizations, or bureaucracies. These, like dinosaurs organizing the protein, coexist uneasily through trading partnerships and symbiosis. While "bureaucracy" has a pejorative connotation in popular argot, this merely reflects and reinforces the opinion of the common man (or common journalist) in his dislike of large formal organizations.

One way of approaching this expanding network of vital interdependencies is to see it as an ecology of social control systems. These systems are necessary for such growth, for they make possible predictability without which interdependence would be short-lived. However, control systems which order the behavior of thousands over wide spans of space and time are, of necessity, bureaucratic in their nature. They take the basic form identified by Max Weber at the turn of the century.

While it is clear that bureaucracy is not necessarily an efficient way of getting work done, (as witness the large and growing literature on its "pathologies"), it is nevertheless an effective means of control. At the heart of the classical empires were the bureaucrats, who maintained order and collected taxes while emperors rose and fell. Such control is critical even though the policies generated may eventually destroy the state or the corporation. For, in an important sense, the control system is the defining characteristic of both.

But what of a society where most are in a large degree burcaucrats or else to a great extent dependent upon them? Although the term "bureaucracy" usually assumes "administration," many of the defining traits are found throughout the ranks of the large-scale organization. Here is Reinhard Bendix's summary of Weber's "ideal type" bureaucracy.

> According to Weber, a bureaucracy established a relation between legally instated authorities and their subordinate officials which is characterized by defined rights and duties, prescribed in written regulations; authority relations between positions, which are ordered systematically; appointment and promotion based on contractural agreements and regulated accordingly; technical training or experience as a formal cóndition of employment; fixed monetary salaries; a strict separation of office and incumbent ... and administrative work as a full-time occupation. [Bendix, 1968]

With the exception of the last condition, unionized labor with guaranteed annual income approaches these conditions, as does the U.S. civil service work force.

In their beginnings most enterprises are led by small, hungry, and entrepreneurial management. This holds for public as well as private enterprise, viz. Sargent Shriver and the Peace Corps, as well as Henry Ford and his horseless carriage. But with success comes increased scale and the pirate ship may be transformed into a fleet of supertankers with an increasing need for rules, records, clear chains of authority and accountability—in short, government (Heller, 1974). Internal order with no direct effect upon the nourishing or threatening environment becomes a dominant concern. Rationalization of work leads to division of labor and requires the integration of the thought and actions of those with special competence and concerns. This in turn results in government by committee, that tedious process which all good bureaucrats decry and live by (Galbraith, 1967). The entrepreneur has become an official.

In response to the problems of increasing scale, the bureaucracy embraces "management science," and tries to use "systems analysis" to define its task and rank its priorities. Planning uses automated information to simulate alternative outcomes—frequently basing the analysis on empirical assumptions less than explicit and empirically grounded observations. As one critic of the modern corporation puts it:

> The system . . . seems a sensible answer to diversity and to problems of maintaining individual responsibility and initiative inside the whale. With the rose-colored spectacles off, the system can be seen as a monumental bureaucracy, with veins made of paper, subject to blood clots at any point. To work at all, the system demands more layers of management, a great burgeoning of staff work, and heavy expenditures of managers' time, simply to cover control, planning, and review. [Heller, 1974]

The author is actually describing one of the many efforts to decentralize large corporations, and his point is that such decentralization itself contributes to further bureaucratization.

As we move down the hierarchy of control, the bureaucracy of the large enterprise is compounded by union protection of the workers' rights or civil service regulations amounting to the same thing. Not only do these agencies contribute to the rigidity of the administrative bureaucracy, they also develop their own bureaucratic rigidity. At the same time, commitment to high technology processes increases interdependency and reduces tolerance for individual deviation and error. It is not an accident that the management of the Hanford Atomic Energy Plant pays close attention to its employees' mental health; the errors that can be made in such a setting are above and beyond the $100,000 misunderstanding. Thus, in addition to regulations flowing from management demands and worker protection, there are also those necessary to carry off complex processes with complex tools. From the ubiquitous computer errors to the electrical blackout of New York City, the results of ignoring these rules are tangible.

In summary: We have a maze of rules growing out of managements' notions of what should be done and how; another growing out of the protection of the workers' rights, and a very basic set growing out of the operations of the organization. It is no wonder that we are made nervous by the *malaise Anglais,* the low rates of growth and productivity in the country where modern industrial society was invented. Between bureaucracy and the unions, the English have created an economy managed by officials rather than entrepreneurs, countered by workers led by strike leaders, rather than labor leaders.

Much of this proliferation of regulations is due to change in the nature of work itself. At one time the technological requirements of work set many limits, from the continuously turning power shafts of the factory to the seasonal requirements of the farm. Today, however, less than one-third of the U.S. labor force works as craftsmen, operatives, or farmers. The rest are professionals, service workers, government workers, and so on. It is an employee society, and, as David Riesman noted long ago, most workers focus not on the hardness of the materials, but on the softness of people and symbols (Riesman et al., 1950).

This change is, of course, due to the automation, first of agriculture, then of industry. The consequent urban concentration of the population has created both new demands for services and the energy surplus to provide them. Some are provided in the private sector; some were once provided in the private marketplace but were taken over by government; some were never provided before and were initiated by government. Some, such as regulation in the public interest, must in their nature be government responsibilities. Thus, a large proportion of the human services is provided by governmental bureaucracies.

THE BUREAUCRATIZATION OF CITY GOVERNMENT

As long ago as 1960 Wallace Sayre and Herbert Kaufman remarked the tendency of the giant service bureaucracies in New York City to strive toward autonomy. With a city government weakened politically by the decline of the parties, and in New York a weakly empowered mayor, each agency struggled for independence.

This was due as much as anything to the sheer size and number of agencies. With over 60,000 workers in the school system and 40,000 in the police department, the inner complexities were hard for any councilman, much less average citizen, to understand or evaluate. How can needs, lacks, and quality of performance be assessed under those conditions? (Sayre and Kaufman, 1960).

Another contributing cause was the power of the civil service regulations. They were originally created to keep the agencies focused on their public duties, rather than the private concerns of politicians and other entrepreneurs—to protect the integrity of the civil servant. They were also intended to guard the principle of hiring and promotion on the basis of merit. Origins do not necessarily predict function, however, and they are increasingly used to protect the welfare of the incumbent civil servants. As Norton Long summarizes the central argument of Sayre and Kaufman:

> The most basic trend in New York has been toward a multi-centered system of increasingly autonomous decisional centers, resistant to integration and wedded to the status quo. The drive for autonomy has been powered largely by the search of the organized bureaucracies and their leaders to escape from the disturbing influence of the "movers and shakers" and to attain the joys of job protection and promotion from within. In this quest, they have been aided by the reformers, the civic groups, and the press. The twin slogans "keep out politics" and "keep out outside interests" have done yeoman service in creating self-contained islands of bureaucratic autonomy in which even the top management posts are reserved for promotion from within. The organized bureaucracies allied to the party leaders have a firm source of support in the Board of Estimate and can balk efforts of the Mayor and his Commissioner to innovate. [Long, 1961]

The system has become even more complicated since Sayre and Kaufman wrote, for the 1960s were a period of rapid growth in the public service unions. "By 1971, more than 25 percent of all state and local governmental employees (2.7 million) were members of employee organizations." As public service employees are three-fourths white collar workers, so were those who unionized, marking a major change in the American labor movement. As one observer put it, the reason was the great boom in human services and the bureaucracies needed to produce and deliver them. The governmental employee "was in the same position as the mass production worker in the 1930s: numerous, needed, and neglected" (Bent and Rossman, 1976).

To be sure, President Kennedy's executive order affirming the right of government workers to collective bargaining had a powerful effect (Rosenbloom, 1971). But this was a canny political judgment of where the society was going; it was going there anyway. The growing strength of unions in helping set prices for items for which the nonunionized must pay was a powerful argument to unionize. So was the growing dependence of the public upon services provided by the bureaucracies and their reactions to strikes and the threat of strikes.

Some observers have seen public service unions as useful counterweights to civil service regulations. They stress the tendency of the bureaucracy to filter messages from below, especially those critical of higher-level performance and policy, and believe union power can guarantee better communication. Having some countervailing power, the argument goes, the lower level workers could expect modification of policy; equally important, they could expect more equity as a grievance system not strictly internal to the civil service bureaucracy became available.

But unions are basically opposed to the merit system, and could erode what merit criteria exists, substituting seniority in promotion. In short, it is possible that the combination of civil service regulations and unionization produces the worst possible output: a property right to the job without respect to merit, and lifetime tenure as well as first choice of higher-level jobs when they open up.

Finally, and not least, insofar as public service unions are successful in seeing that their members obtain parity with private market union members, the result must be increases in salaries. The resulting tax burdens can be very destructive, leading even to the city government's defaulting on its debts and going bankrupt—thus losing the right to govern itself. Furthermore, we can expect strikes of workers supplying critical services to the city; without such strike power unionization has only the strength of persuasion. In short, public service unions combined with civil service regulations promise the emerging city higher-priced services with no improvement in quality, higher taxes with no public benefit [Nigro, 1976]. So dependent is the citizen, however, on these human services, that there may be little remedy at the municipal level; it may have to come from the larger society.

Remedies from the state and national governmental levels take two forms: transfers of money and regulation of programs. The two are not always identical, for while categorical grants continue to be very important, revenue sharing has introduced a new dimension to intergovernmental relations. The two strategies produce complementary vices. Categorical grants, for example those funding mental health centers or health planning, tend to produce rigid definitions of the means and fuzzy definitions of the goals. They aim for national, or at least regional, uniformity. They result in confusion, distortion of purposes, and patterned evasion of the norms by those who must make programs work in Boise and Baton Rouge, as well as in midtown Manhattan.

But the doling out to states of federally collected tax monies also has its vicious side. Revenue sharing is essentially a scheme devised by presidential adviser Walter Heller to dispose of an expected federal budget surplus; by the time it was enacted there was a deficit in Washington and revenue sharing contributes substantially to it. Nowadays it is a device by which the federal government collects the taxes in its own name, accepting the onus, and passes them with practically no strings attached to the state and local governments. Local officials accept the "federal money," while opposing it in principle, thus biting with impunity the hand which feeds them (Banfield, 1976). The net result is to relieve the city government of responsibility for both taxation and new programs. It makes easier the persistence of existing arrangements (Pressman, 1975).

Federal funding of categorical programs, from mosquito abatement to health services, is accompanied by regulations which increase in range and specificity as the program burgeons. The original Social Service Act, for example, was exactly one paragraph long. As the program encountered various threats and challenges, additional clauses were added, until today the act as amended, plus regulations for implementing it by bureaucracy, is a thick tome. Such complexes of rules are meant to maintain the integrity of the act as Congress intended it to be; the result is often to so snarl up the work of the local agency as to require a special "office of foreign affairs" to deal with

Washington. In the federal-local relationship the bureaucracies grow in size at both ends of the line.

A further type of federal regulation which the city agencies must abide by is that which is intended to protect citizens' rights. The most important of these is probably the Equal Opportunities Act, which bars discrimination on the basis of most demographic characteristics in hiring and promoting workers and in providing services. Since it is a law, city agencies must pay close attention, for they can lose federal funding on the one hand, and be sued by employees or clients on the other.

A third type of regulation from the federal government began with the "maximum feasible participation" phrase in the first Urban Renewal Act of 1954. While it had little effect on the operations of local public authorities using urban renewal, the principle of citizen participation in the governing of such local operations as neighborhood health centers and "model city" programs became widespread during the 1960s and 1970s. From the vague admonition in the Urban Renewal Act, the requirement became more specific until, in the Community Mental Health Centers Act, the governing boards of centers must reflect the nature of the population it is supposed to serve. This means the proper percentage of white and black (and assorted others), men and women, the young and the old, must be represented on the board in proportion to their numbers in the community; furthermore, 85% of them must live in the area served.

The results of these regulations are typically transmitted from the city agency to the relevant Washington agency. However faithfully they are carried out, they have one predictable result: Paper, paper, and more paper, written in the curious, vacuous prose of bureaucratize. Naturally, they require someone to write them, to count heads, to fill in forms, after studying the regulation intensely to see what kinds of verbal compliance are best (a) to satisfy the interest of whoever might read it, and (b) to interfere as little as possible with the way the city agency prefers to do its business. Writing such reports and proposals is a relatively new, but rapidly growing craft among all bureaucracies operating on an "entitlement" economy.

A second result is either the loss of some of the city agency's freedom to do the best it can given the local situation, *or* the aforementioned "patterned evasion of norms." And since most programs are at least partially funded by the city, county, or state, the local administrators have some legitimate resentments, especially if the regulations are costly to resources and program operations. As one harassed bureaucrat remarked: "The Feds only support thirty-five percent of the program, but they want one hundred percent control" (Greer, 1978). Some agencies have been known to refuse federal funding for these reasons, and some administrators to resign. The entitlement economy replaces the market with a cumbersome, expensive, and only partially effective inside review.

The coming of the computer has been a mixed blessing to the bureaucracies of the city. Its value is greatest in handling routine data such as payrolls material costs, and the like. Here it quickly, and usually accurately, provides the running tallies necessary for oversight and control. Its value is usually nil or worse when it is used for policy-making. "Operations research" is chiefly a computer-programming skill, since its practitioners are usually without discipline or subject matter expertise. It works best when the assumptions are correct, e.g., when they are based on empirically grounded theory, and when the problem is subject to measurement adequate to the mathematical operations; neither obtained in the late Vietnam War, and as a result, the term "body count" has achieved a hateful immortality. Like police officers required to fill a quota, American military bureaucrats knew how to write a ticket.

Usually, however, computer-based thinking does not gravely harm policy-making and administration because it is vague or empty of policy implications. What it does harm is self-criticism in the bureaucracy, and dininishing storage space in the file drawers. In its formal structure and conceptual emptiness, it unfortunately complements perfectly some of the worst vices of the bureaucratic mind. It is directly analagous to accounting, as Heller views it:

> The accountant is playing an elaborate game, whose purpose is not to tell the truth (for truth is a chimera), but to obey the rules. The rules in turn are not designed to make managers more efficient, or to inconvenience crooks, or even to keep the company solvent. They are there simply to allow the game itself to proceed. That explains how accountants can usually wash their hands of blame or dirt when disaster strikes. They can nearly always prove that they played their game according to their rules. Unfortunately, management is a different game entirely. [Heller, 1974]

Running a city agency is a very different game from running a computer program.

While some of the city's business is with bricks and mortar, tangibles which can be seen and evaluated against the private sector, the human services agencies (health, welfare, education, and the like), are another matter. They usually cannot be measured against the private sector, often because parallel facilities do not exist, or, if they do, operate at a much smaller scale or for different ends. (Parochial schools, for example, are part of a large-scale system, but they are usually underfunded, and their major reason for being is the propagation of the true faith.)

How then do you evaluate a bureaucracy when there is no bottom line? Elsewhere I have put it this way:

> We use bureaucratic strategies on the bureaucrats. In doing so, however, we find that every strategy has serious flaws. We try to evaluate the quality of service by structure, by process, and by output. In the structural approach, we ask: Are the necessary resources present in sufficient numbers? The buildings, materials, machinery, and bodies with appropriate certificates? In the processual evaluation we ask: Did the various actors do all they were supposed to, without wasting resources on the superfluous? Did they give the test, check the statute, handle the legal cases on time, consider and act upon the bills before them? And if they did—still the question: What difference? The third strategy of accountability holds the actor responsible for the *outcome* of his actions. This is the most critical matter of all: only from gauging outcomes can we measure what the agency produced. Only as it relates *to* outcome does either structure or process standard have validity. Yet to evaluate outcome is the most difficult type of evaluation. We must know exactly what we expect; and control for a host of variables which affect the difficulty of producing it. [Greer et al., 1978]

Efforts at evaluation are, thus, typically structural or processual. The first produces endless inventories, the second endless rules. Both tend to further ritualize the ritualized. Indeed, they frequently generate far more data than can be understood, for if there are rules of importance, it is important to measure compliance. Yet so august an agency as the Internal Revenue Service matches only 6% of the withholding forms from employers with the individual income tax report (Academic Information Service, 1977).

If one could evaluate the outcomes of bureaucracies, it would be possible to relate the individual's contribution to that outcome and reward it accordingly. Failing such measures, most agencies fall back upon licensure (years of graduate work for teachers, board certification for physicians), number of employees supervised (however badly), that old standby, seniority (or gerontocracy), and the like. Naturally, there are enormous discrepancies between reward and contribution.

Some have even raised the question: Do organizations have goals? To paraphrase Charles Perrow: Individuals have goals, but organizations only have usages—different usages for different groups and individuals (Perrow, 1972). Thus the school system has one set of uses for the administrators, another for the teachers, another for the parents, and so on. This is a beguiling approach, and there is a hard core of truth: All bureaucracies tend to be seen as *cosa nostra* by the incumbent bureaucrats. But looking at city governmental agencies, it is clear that they have assigned tasks arising out of the polity. One may be cynical about whether the agency "works," whether it does have the effect the political leaders intended when it was created, but there *were* expectations at city hall, the state house, or on the Hill.

THE DOCTRINE OF ENTITLEMENT

As the party system and the urban political machine weaken, the polity of the city tends to be more and more based on single-issue interest groups. One important issue tends to be security in the job and a reward that is felt to be just. (The fact that the two are contradictory seldom troubles us.) The struggle for security is rapidly turning a politics of pluralism into a complex of mangers occupied by a variety of dogs. The civil service, union protection, and tenure in all its varied forms, seem to be an inescapable concomitant of the welfare state; all are income maintenance schemes. They are all based on "equity claims," which as Morris Janowitz notes, "are the income and services that a person and the members of his family come to hold as accruing to him regardless of ownership of the capital resources involved" (Janowitz, 1976).

Such claims, first embodied in the phrase "the property right to a job," make the selection and use of a labor force increasingly inflexible. They force the retention of dead cells, which must be wired around if the machinery is to run. When one considers that the claims, extended through fringe benefits, are increasingly expensive, then scarcity of resources is guaranteed and so is competition among the bureaucracies.

It is no wonder that the bureaucrat becomes a lobbyist for his client—the client is the basis for his own equity claims.

The *clients* also fight to keep and extend their equity claims, and these claims are not limited to those on welfare.

> These programs penetrate well beyond the lowest social stratum. Thus, for example, over 100,000 members of the armed forces in the early 1970s were eligible to receive food stamps. Old-age insurance has become a crucial device for members of the middle strata to relieve themselves of, or shift the burden of caring for their aging parents. Veterans' benefits are social welfare, since they are as much designed to overcome liabilities as they are to be rewards. The emerging programs of medical insurance are more and more broadly based. Significantly, elements of these programs are closely linked to the cost of living index. [Janowitz, 1976]

While confrontation politics still breaks out from time to time, a much more potent weapon of the client is litigation. The "class action suit" has become a household word, and litigious politics contributes to the dangerous erosion of trust which originated in the Vietnam years. "Entitlement" ceases to have an aura of *noblesse oblige* and instead accrues connotations of rancor, resentment, and self-righteousness.

The seeds of many equity claims were planted during the depression of the 1930s, but since 1945 entitlement has spread rapidly to embrace more types of advantage, and types of people, at an enormously increased cost. This came about partly because of the rapid and steady increase in productivity and wealth, partly because of naiveté. The governments responded to perceived need and the tradition of giving, paying little attention to costs (we could afford them) or benefits derived at given costs (they were assumed to be worthwhile because needed). Now that economic growth has slowed to a terrapin's pace while the costs of human services continue to grow, the bureaucratized city faces the serious probability of forced retrenchment. Given the multitudinous equity claims and the constituencies based upon them, it will not be an easy or pleasant operation. Unemployed or underrewarded white-collar workers and would-be or have-been bureaucrats are a notorious source of political instability. We may be forced to continue the public bureaucracies as welfare agencies for such a population.

To repeat: Bureaucracy may or may not be an efficient productive system, but it *is* an effective instrument of social control. The Nazi party could never have devised a means for governing Germany, but in the bureaucracy of the Prussian Empire and those of the various petty states, an instrument stood ready. Good, rule-ridden Germans operated the state for Hitler. So might the crisis of a no-growth, stagnating economy be handled in the United States, with control of the people through control of bureaucratic heads and their empires, and the distribution of scarce resources through the same instruments.

REFERENCES

Academic Information Service, Inc. (1977) 1978 Tax Guide for College Teachers. Washington, DC: Ch. 17. Ture for Typewritten W-2 Forms. Magnetic tape data was cross-filed.
BANFIELD, E. C. (1976) "Revenue sharing in theory and practice," pp. 88-99 in Urban Administration. Port Washington, NY: Kennikat, 1976.
BENDIX, R. (1968) "Bureaucracy," in International Encyclopedia of the Social Sciences. New York: Macmillan, 1968.
BENT, A. E. and R. A. ROSSMAN (1976) "Management, politics and change," in Urban Administration. Port Washington, NY: Kennikat.
GALBRAITH, J. K. (1967) The New Industrial State. New York: Mentor.
GREER, S. (1978) Current field research on Community Mental Health Centers
———, R. D. HEDLUND, and J. GIBSON (1978) "Introduction, the accountability of institutions in urban society," in Accountability in Urban Society, Vol. 15, Urban Affairs Annual Reviews. Beverly Hills: Sage Publications.
——— (1972) The Urbane View. New York: Oxford University Press.
——— (1962a) The Emerging City. New York: Free Press.
——— (1962b) Governing the Metropolis. New York: John Wiley.
——— (1955) Social Organization. New York: Random House.
HELLER, R., (1974) The Great Executive Dream. New York: Dell.
JANOWITZ, M. (1976) Social Control of the Welfare State. New York: Elsevier.
LONG, N. (1961) "Sayre and Kaufman's New York: competition without chaos." Public Admin. Rev. 21 (1).
NACHMIAS, D. and D. H. ROSENBLOOM (forthcoming) Bureaucratic Government: An Approach to American Politics. New York: St. Martin's.
NIGRO, F. A. (1976) "The implications for public administration," in Urban Administration. Port Washington, NY: Kennikat.
PERROW, C. (1972) Complex Organizations: A Critical Essay, Glenview, IL: Scott, Foresman.
PRESSMAN, J. L. (1975) Federal Programs and City Politics. Berkeley: University of California Press.
RIESMAN, D., R. DENNY, and N. GLAZER (1950) The Lonely Crowd. New Haven, CT: Yale University Press.

ROSENBLOOM, D. H. (1971) "The development of the public employment relationship," in Federal Service and the Constitution, Ithaca, NY: Cornell University Press.

SAYRE, W. and H. KAUFMAN (1960) Governing New York City. New York: Russell Sage Foundation.

Part II

The Tax Revolt and Policy Priorities

5

Citizen Preferences and Urban Policy Types

WAYNE LEE HOFFMAN
TERRY NICHOLS CLARK

☐ DECLINES OR SLOW GROWTH IN THE URBAN PUBLIC sector will increasingly require cutbacks in some services if other areas within the public economy are to expand. Furthermore, since preferences among urban groups differ, shifts in the service mix have implications for who will be attracted to or repelled from the city. This chapter distinguishes among various types of urban services and identifies the economic groups that will tend to support or oppose particular service categories.

THEORIES OF POLICY OUTPUTS

This paper addresses four questions, each emerging from a somewhat distinct body of previous work.

(1) *Do citizens have any structured preferences for urban public policies?* Most debate about the structure of policy preferences has

AUTHORS' NOTE: *This is research report #85 of the Comparative Study of Community Decision-Making. We are grateful for financial support from USPHS, NICHD, HD08916-03. Opinions expressed herein in no way represent those of the Urban Institute or agencies which supported its research.*

concerned national policies. The view of Converse (1964) and others that minimal constraints exist among citizen policy preferences has been challenged (e.g., Nie, Verba, and Petrocik, 1976; Page, 1977), but almost never with data on urban policy preferences. Here we address this issue using the most extensive study to date of urban policy preferences of American citizens: the Urban Observatory study of 4,266 citizens in ten cities. Do we find constraints among citizen preferences? To anticipate our results: some.

(2) *Is there a distinguishable left-right continuum among urban policy preferences?* Many have sought to distinguish left-right continua for national policies using various types of data, but little work has been undertaken on such issues at the local level. Much empirical work on cities has built on neo-Downsian assumptions in positing a left-right continuum, but analyzed only aggregate city-level data, such as income and municipal expenditures (e.g., Davis and Haines, 1966; Bergstrom and Goodman, 1973). Our answer in brief: Yes, we can distinguish a left-right continuum using individual-level data concerning urban services.

(3) *Is there a distinguishable public good-separate good continuum in urban policy preferences?* The public good concept has played an important role in theorizing about public policy since Samuelson's (1954) formulation. But most conceptual as well as empirical analyses have proceeded as if only two types of policies exist, pure public goods and pure private goods (e.g., Bradford, 1971; Bergstrom and Goodman, 1973). One of our basic arguments is that policy analyses can be sharpened by using a continuum from public to separable goods. Public policies can then be ranked on such a continuum and differentially related to political system characteristics (Clark, 1973: 59-67). Similarly, different social groups differentially prefer more public or more separable goods, and find differentially legitimate the operation of political systems based on each (Clark, 1975). But as almost all work to date on this public-separable continuum of public policies has used city-level data, can we distinguish such a continuum using citizen survey data? In brief: yes, although in these data it overlaps considerably with the left-right continuum.

(4) *How can such citizen policy preferences be explained?* Two general types of interpretations are considered. First is the *utilitarian view* that citizens seek to maximize their material well-being. Our model is straightforward. In the case of urban public services, where citizens are taxed, generally in proportion to property value, the rich

pay more per capita than do the poor. Even if we assume that local taxes are neither progressive nor regressive, but take an equal percentage of one's property value at all levels, the affluent are likely to be less satisfied with their tax/service exchange than the poor.[1] This is because in the private market both rich and poor convert a dollar of income into a dollar of consumption. But at least for pure public goods, a high or low tax bill yields identical public services. However, urban public services differ both in the degree to which they are available equally to all citizens, (i.e., in separableness), and to which they redistribute income from rich to poor. The more separable a good, the lower the cost of private sector alternatives to those who seek them: for example, public hospitals can be ignored by those who prefer to pay for private hospitals. But clean air—a quite public good—is virtually unobtainable for one or a few urban citizens; it is normally provided to all or none.

These considerations lead us to hypothesize:

(A) *The higher the income of the citizen, the lower his preference for urban taxes and services.*
(B) *The more redistributive the service, the more strongly will proposition A hold.*
(C) *The more separable (and less public) the good, the more strongly will proposition A hold.*[2]

Propositions A, B and C formally imply:

$$P_i = A + \delta I + \beta S_i + \gamma R_i, \qquad [1]$$

where

P_i = citizen's preferred level of urban public service i
I = income level of citizen
S = degree of separableness of i (high = very separable, low = very public)
R = degree of redistributiveness of i (high = very redistributive)
δ, β, γ = coefficients, all hypothesized to be *negative*. A = constant.

A second view suggests a *curvilinear* relationship between preference and income. Banfield and Wilson (1963: 35-36) have argued that urban "conflict is generally between an alliance of the low-income and high-income groups on the one side and the middle-income groups on the other.... The middle-income group generally wants a low level of public expenditures." Upper-income persons, they argue, share a preference for high public expenditures with low-income persons because higher-income persons are more "public regarding." This argument was refined in their later works (1964, 1971), although they downplayed the curvilinear aspect of the hypothesis. Wilson (1975) subsequently suggested that middle-income persons have become particularly dissatisfied with governmental activities since the late 1960s, even if they may obtain substantial benefits from government.

The converse argument was made by Stigler (1970). Extending work by Aaron Director, he developed a simple model which holds that middle-income persons will benefit *most* from government spending because they are able to create coalitions with the largest voting base. A related if less elegant argument was made by Phillips (1970). Weicher (1971) tested Stigler's hypothesis using police patrol data for Chicago neighborhoods, and found exactly the opposite: Middle-income residents paid more in taxes for the level of police service than the poor or rich.

While specific theories and empirical results of these authors differ considerably, they agree on the curvilinear nature of the income-spending preference relationship. They lead us to three specific hypotheses which parallel the utilitarian-inspired hypotheses above.

(D) *Income is related to preferences for urban public services in curvilinear manner: low and high income citizens prefer higher service levels, middle income citizens prefer lower service levels.*
(E) *The same curvilinear relation should hold for more redistributive services.*
(F) *The same curvilinear relation should hold for more separable (and less public) goods.*

In equation [1], β, and γ should thus be lower for medium-income levels than for low- or high-income levels.

Most results below support the three utilitarian hypotheses more than the three curvilinear hypotheses.

In addressing our fourth question—how can preferences be explained—an addition to these propositions is useful. Since citizen evaluations of services are related not only to initial preferences, but also to experience with service provision, spatial characteristics of policies are likely to interact with other explanatory factors. Specifically,

(G) *The more public the good, the more citizen satisfaction with it is explained by city-level (in contrast to individual- or neighborhood-level) characteristics.*

The logic here is that different separable goods within the same city will be consumed by citizens differing in individual and neighborhood characteristics. For example, police protection may vary across neighborhoods. As service patterns are likely to differ more for such separable goods within a city, and citizen preferences, in turn, to differ in conjunction with such intracity variations, city-level characteristics should be less important in explaining citizen satisfaction than for more public goods. More formally, consider

$$P_i = a_i R + b_i N + c_i C + e_i \qquad [2]$$

and

$$P_i = f_i R + g_i N + h_i, \qquad [2a]$$

where

P_i = preference of respondent for spending on service i
R = individual respondent's characteristics
N = respondent's neighborhood characteristics
C = respondent's city characteristics
a_i, b_i, c_i, f_i, g_i = coefficients to be estimated
e_i, h_i = residual error terms

Then, Proposition G implies

$$c_p/a_p > c_s/a_s, \qquad [3]$$

where coefficients c and a are defined in [2] and the subscripts p and s designate public and separable goods respectively. A slightly different statement of the same idea will be tested below:

$$R^2_{C_p}/R^2_{R+N_p} > R^2_{C_s}/R^2_{R+N_s}, \qquad [4]$$

where

$R^2_{C_p}$ = Net increase in R^2 due to city characteristics = R^2 for regression estimate of equation [2] - R^2 for regression estimate of equation [2a]. (Subscripts: C = city and p = public good.)

$R^2_{R+N_p}$ = R^2 for regression estimate of equation [2a]. (Subscripts: R = respondent, N = neighborhood, and p - public good.)

The two terms on the right-hand side of expression [4] differ only in that the subscript s = separable goods.

We see no compelling theoretical reasons for hypotheses analogous to G for policies varying in redistributiveness.

THE STUDY

In the study a representative sample of citizens in each of ten American cities was interviewed about urban policy issues (total usable N = 4,266.) The survey was part of the Urban Observatory program, supported largely by HUD, involving local public officials and academics in each of the ten cities. Following urban disorders and efforts to enhance citizen participation in the late 1960s, the study posed questions about urban policies most critical to American cities at the time. When fieldwork was undertaken in the summer of 1970, citizens had probably experienced more efforts to rouse and inform them than at any other time in recent history. As the largest and most systematic effort of its sort ever undertaken, the study is particularly well suited to address the types of hypotheses listed above. The timing of the study probably weights it toward capturing coherent attitudes; but only additional surveys can document changes. Further details on the survey are reported in Fowler (1974); we added neighborhood and city data as described below.[3]

Our analysis focuses on one basic questionnaire item. It was presented to the citizen with a list of policy areas:

> Here is a list of services and policies. Some we have talked about already, others we have not. For each I want you to tell me whether you think the local agencies should spend *more* money, less money, or about as much money as is now spent on those services or problems. Remember, that to spend more on something, the local government either has to spend less on something else or it has to raise taxes.

Policy areas listed were: public schools, police patrolling the streets, street lighting, cleaning and repairing streets, providing medical aid to people who cannot afford to pay for it themselves, trash and garbage collection, building low-cost housing, controlling air pollution, improving public transportation, welfare, and Aid for Dependent Children (AFDC).[4]

PREFERENCE STRUCTURES: LEFT-RIGHT AND PUBLIC-SEPARABLE GOODS CONTINUA

Policies may be classified in terms of their potential for redistribution of income from more to less affluent citizens. This is perhaps the most fundamental aspect of a left-right continuum. Without specific details on production and consumption of services in each city, we cannot be certain of the amount of redistribution they involve. More important for present purposes is how policies are perceived by citizens. Table 1 lists the ten services simply in terms of "low," "higher," and "high" potential for redistribution.

·Second, policies may be classified in terms of their separableness, also summarized in Table 4. The most public goods are those supplied

TABLE 4: CLASSIFICATION OF SELECTED LOCAL GOVERNMENT SERVICES ACCORDING TO TWO DIMENSIONS OF POLICY EFFECTS

Name of Policy Cluster	Example of Local Service	Potential for Redistribution	Degree of Separableness
Environmental services	1. Controlling Pollution	Low	Citywide
	2. Mass transit	Low	Citywide (with some neighborhood)
Traditional services	3. Police protection	Low	Neighborhood or block level
	4. Street lighting	Low	
	5. Street repair	Low	
	6. Trash collection	Low	
Public Schools	7. Public schools	Higher	Neighborhood
Welfare services	8. Public housing	High	Neighborhood and Individual
	9. Medical care	High	Individual
	10. Public assistance	High	Individual

TABLE 5: CORRELATIONS AMONG TEN LOCAL POLICY ITEMS[a]

	Control Pollution	Public Transit	Police	Street Lighting	Street Repair	Trash Collection	Public Schools	Public Housing	Medical Care	Public Assistance
Control pollution	—	.34	.17	.14	.10	.12	.14	.24	.13	.09
Public transit	$\bar{X} =$	—	.21	.23	.16	.23	.15	.31	.22	.19
	.34									
Police			—	.34	.29	.27	.14	.14	.17	.13
Street lighting				—	.48	.39	.10	.26	.21	.18
Street repairs				$\bar{X} =$	—	.52	.24	.25	.27	.34
Trash collection				.38		—	.30	.31	.34	.34
Public schools						$\bar{X} = .29$	—	.28	.33	.32
Public housing								—	.52	.50
Medical care									—	.60
Public assistance									$\bar{X} = .54$	—

[a] Correlations are tau gamma's.

in a broadly similar manner throughout the city. They may not be consumed equally by all citizens, but are at least generally available and not readily reallocable on social or geographic criteria. If they were pure public goods, they would be indivisible and costs of excluding citizens from consuming them would be very high. Pollution control is the most public good in this sense; mass transit just slightly less so. By contrast, the traditional city services of police, street lighting, street repair, and trash collection are largely neighborhood- or block-specific, both for technical and administrative reasons. Finally, the most separable services are provided to individuals or families, often on the basis of demographic and financial criteria which they must meet to receive

TABLE 6: ROTATED (CORRELATED) FACTOR PATTERN FOR PREFERENCE ITEMS[a]

	Welfare Services	Traditional Services	Environment
Control pollution	-.01	-.06	.80
Public transit	.15	.04	.68
Police	-.16	.49	.25
Street lighting	-.04	.69	.04
Street repairs	.06	.73	-.14
Trash collection	.18	.60	.07
Public housing	.77	.00	.20
Medical care	.68	.01	.00
Public assistance	.77	.03	-.05

Correlations Among Three Rotated Factors

	I	II	III
Welfare services	—	.19	.09
Traditional services		—	.18
Environment			—

	Eigenvalues for Nine Item Solution	Percentage of Total Variance Attributed to Each Factor
Factor 1.	2.25	25%
Factor 2.	1.27	14%
Factor 3.	1.11	12%
Factor (4.)	.78	(10%)
		52% 3 Factor Total

[a] Eigenvalue for public schools is below 1.0 and thus is not shown.

them: public assistance, public hospitals, and public housing. Because of external effects on the neighborhood produced by public hospitals and public housing, they seem slightly less separable than public assistance.

The intercorrelation matrix of responses to the ten policy items appears in Table 5. Boxed are policies conceptually grouped along the two continua in Table 4. The higher within-cluster correlations lend support to the conceptual ordering. Similar if more striking results emerge in factor analysis. Principal components, varimax, and oblique rotations were computed, all generating similar solutions. Table 6 displays the factor pattern from the oblique solution. The fourth factor generated a loading only for public schools, not shown as its eigenvalue fell below 1.0. The four types nicely parallel those in Table 4. Our three initial questions all seem answered in the affirmative by these results.

HOW TO EXPLAIN THE SOURCES OF CITIZEN POLICY PREFERENCES?

The three utilitarian and three curvilinear hypotheses can be simultaneously assessed in Figure 2, which displays relationships between income and responses to the four types of policy questions as well as to a composite measure. The Overall Preference measure is composed of all ten items based on a factor scale.[5] The results are generally consistent with the three utilitarian hypotheses: income is most negatively related to the highly redistributive welfare policy. Traditional services, less redistributive and less separable, show a weaker relationship. No consistent relationship holds for schools and environmental services. The major negative relationships are all monotonic rather than curvilinear.[6]

Still, Figure 2 shows only bivariate relationships; these may be affected by other characteristics of the individual citizen, his neighborhood or city. Table 7 shows the results of ordinary least squares regressions computed to estimate equation [2a]. The low R^2s may lead some readers to feel that the answer to our question no. 1 is that citizens have only slightly structured preferences: combined individual, neighborhood, and city-level effects leave 82 to 94% of the variance in the policy items unexplained.[7] Still, the sample is so large that not only are all full equations highly significant, so are many individual variables.

Proceeding down the list of independent variables in Table 7, we see that either *income* or one of the *socioeconomic status* measures remained significantly related to spending preferences in the multivariate analysis—and interaction effects of income with type of policy issue recur similar to those in Figure 2. Separate analyses were undertaken of the three socioeconomic status variables (income, education, and occupation). When all three were additive with the same sign, only the combined SES measure was included. The more affluent generally prefer fewer public services, especially welfare-type services. However, when education rather than income is considered, the more highly

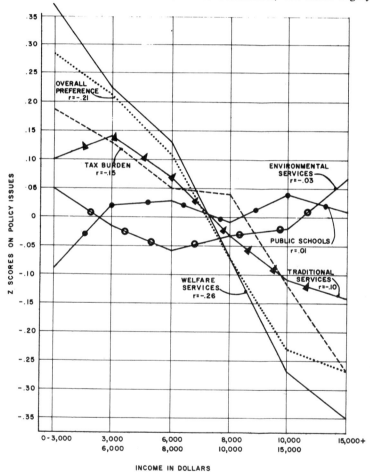

FIGURE 2: SPENDING PREFERENCE FOR SIX POLICY ISSUES BY INCOME LEVEL

TABLE 7: BETA COEFFICIENTS FOR VARIABLES IN EXPANDED MULTIPLE REGRESSION MODEL (N-3999; significance at .05 level, two-tailed test, indicated by*)

Independent Variables	Overall Preferences	Dependent Variables			
		Environmental Services	Traditional Services	Public Schools	Welfare Services
R^2	.148	.048	.083	.095	.165
R^2 Adjusted	.144	.044	.079	.090	.161
Significance level	.001	.001	.001	.001	.001
Respondent Characteristics					
Combined SES	-.06*	N.A.	-.04*	N.A.	-.14*
Income (log)	N.A.	-.02	N.A.	-.01	N.A.
Education	N.A.	.05*	N.A.	.09*	N.A.
Occupational prestige	N.A.	.04*	N.A.	.03	N.A.
Black	.25*	-.05*	.20*	.17*	.23*
Spanish	.08*	-.01	.07*	.01	.08*
Catholic	-.01	.02	.00	-.02	.00
Jewish	.06*	.04*	.03*	.04*	.05*
Age	-.02	.09*	.04*	-.14*	-.04*
Child present	-.02	-.02	.00	N.A.	-.04*
In private school	N.A.	N.A.	N.A.	.00	N.A.
In public school	N.A.	N.A.	N.A.	.03*	N.A.
Years in city	.00	-.03	.00	.00	.02
Size of birthplace	-.01	.02	-.06*	.03*	.02
Homeownership	.07*	.01	.01	.09*	.11*
Housing quality	-.02	.03	.01	.02	-.05*
On welfare	N.A.	N.A.	N.A.	N.A.	.04*

TABLE 7 (cont.)

Independent Variables	Overall Preferences	Environmental Services	Traditional Services	Public Schools	Welfare Services
Lives in public housing	N.A.	N.A.	N.A.	N.A.	.03
City workers	-.00	-.01	.01	-.01	.00
Perceived Service Distribution					
Service(s) better	.06*	.06*	.02	.04*	.04*
Service(s) worse	.09*	.03	.12*	.02	.04*
Neighborhood Characteristics					
Percentage renters	-.01	.07*	.06*	.00	.03
Percentage vacant	.07*	-.02	.08*	.05*	.05*
Percentage crowded	-.03	-.05*	-.01	-.04*	-.01
Percentage in large structure	.01	.06*	.01	-.05*	-.02*
City addition to R²	.035	.024	.043	.028	.020
Significance level of joint city distribution	.001	.001	.001	.001	.001

Variables Used in Expanded Models

Combined SES   Sum of Z-scores of interval measures of annual family income, respondent education, and head of family's occupational prestige score.

Occupational prestige   Duncan occupational prestige score coded for primary occupation of family head.

Income   When used alone, the logarithm to base 10 for annual family income in thousands.

Education   Respondent's highest year of education.

TABLE 7 (cont.)

Homeownership	Dummy variable coded 1 if family rents and 0 if owning or buying unit where interviewed.
Black	Coded 1 if respondent is black, 0 if otherwise.
Spanish	Coded 1 if respondent is Spanish-speaking American, 0 if otherwise.
Age	Respondent's age in years; mean 42.7.
Child present	Coded 1 if respondent's household included one or more children under the age of 18, 0 if not.
In public school	Coded 1 if children present had or would attend public schools in the city; 0 if not.
In private school	Coded 1 if children present had or would attend private or parochial school; 0 if not.
City worker	Dummy variable coded "1" if respondent or any "close relative" works for the city government; "0" if answer is no.
Catholic	Coded 1 if respondent was raised a Catholic; 0 if not.
Jewish	Coded 1 if respondent was raised as a Jew; 0 if not.
Years in city	Number of years respondent lived in the city of interview.
Size of birthplace	Coded 1 if respondent was born in town of less than 10,000 in population, on a farm, or in "the country" unspecified; 2 if born in SMSA of 650,000 or less in 1960; coded 3 if born in SMSA of 650,000+ in 1960.
Housing quality	Interviewer's rating of respondent's dwelling unit; coded 1 if substandard, 2 if "deteriorating," 3 if sound.
On welfare	Coded 1 if respondent's family received any income during 1969 from AFDC or general relief; 0 if not.
Public housing	Coded 1 if respondent lives in public housing; 0 if not.

TABLE 7 (cont.)

Perceived distribution variables — Two dummy variables: the first coded 1 if respondent said neighborhood services were better than in the rest of the city, 0 if not; and the second coded 1 if services were reported as "not as good" as in other areas of the city, 0 if this was not the response. The items included ten basic neighborhood services. For public schools preferences, quality of schools comparison was coded in similar fashion.

Neighborhood Characteristics

The following four variables are derived from 1970 Census public use data with percentages based on total units in the respondents' "neighborhood" (zip code area).

Percentage renters — Percentage of neighborhood households who live in rented quarters.

Percentage vacant — Percentage of neighborhood dwelling units that have been vacant 6 months or longer.

Percentage crowded — Percentage of all dwelling units that have an average of more than one occupant per room.

Percentage in large structures — Percentage of all dwelling units that are in apartment buildings with 5 units or more.

City

Coded 1 if respondent lives in the city, 0 otherwise.

There are no missing values in the independent variables. Missing cases for certain dependent variables reduced the N to 3,999. For all dummy variables, original missing values are included in the 0 category. For some interval level variables, the mean was substituted for the missing value. For other interval variables (particularly the three SES measures) an estimate from closely related variables based on simple regression were substituted for the missing values. Further details on most items are in Fowler (1974).

Where a variable was not included in the equation, usually because of a substitution of a similar variable, "N.A." appears.

To assess interactions of the variables with individual cities, the equations used for Table 7 were reestimated separately for each of the 10 cities. All major results held consistently within each city, except for a few instances where the number of respondents in some categories fell too low to permit analysis.

educated prefer more spending for public schools and for environmental services (pollution control and public transit).

Considering next *ethnic and religious variables, Black* citizens were the most decisive in their responses, consistently favoring more spending in all areas except environmental services.[8] Spanish-speaking citizens responded in similar manner to Blacks, if less consistently. Jews (3% of the sample) also showed the same response pattern, except that they favored more spending on environmental services. Catholic-non-Catholic differences were minimal. Combined religion and national origin variables were also analyzed, such as Irish Catholics, German Protestants, and so forth, but most results were insignificant.[9]

Older citizens tended to favor environmental and traditional services, but opposed school or welfare increases. Parents with a child in public schools favored increased school expenditures; having a child in private school was unrelated to spending preferences. Length of residence in the city was included to see if through greater local attachments it might increase demand for public services; we found no such effects. Size of birthplace was used to consider the hypothesis that persons from rural and small town areas would prefer less spending; this effect was only present for traditional services, and ran in the other direction for public schools. Homeowners stood out as favoring less spending overall, and particularly less for schools and welfare services. This is in part explained by the greater visibility of property taxes to homeowners than to renters, who perceive local taxes and service costs less than homeowners, even though they may pay as much or more indirectly (Hoffman, 1975; Aaron, 1975). Housing quality (an interviewer judgment) was included to assess any positive relationship between private sector consumption and public service demands. The only significant result was for welfare services, which persons living in poorer housing favored increasing. Persons receiving welfare benefits also favored more welfare spending, although living in public housing had no additional effect.

Citizens working or having a close relative working for the city government ("city worker"), perhaps surprisingly, showed no distinct preferences for increased spending.

Citizens who reported that their neighborhoods received either better or worse public services than other neighborhoods generally favored increased spending. More detailed analysis of these and related items suggested that they indicate more of a concern with public services generally than an "objective" assessment of services received.

(For example, Blacks tended to report that their neighborhoods received poor service delivery much more often than whites living in the same neighborhoods.)

However, even if self-reports on service delivery express respondents' perceptions, they may also be related to certain service characteristics. In any case, inclusion of the two variables in the regression is one step toward removing neighborhood effects which could distort coefficients of other variables. Another step is taken, again imperfectly to be sure, by adding four basic census characteristics for respondents' neighborhoods.

Citizens in neighborhoods with a high percentage of renters favored more spending for environmental and traditional services. Those in neighborhoods with high vacancy rates showed the same pattern as poor and Black citizens: They favored more traditional services, schools, and welfare, but not environmental services. Residents of neighborhoods with "crowded" housing preferred less spending for environmental services and schools, while those in neighborhoods with a high percentage of residents in large structures favored more environmental services and less spending on schools. These effects may have some substantive interest, but like perceived service quality, they are included here also to eliminate their effects on other variables in the model.

Finally, after removing the variance due to these many respondent and neighborhood characteristics, do any net *city effects* remain? This question was addressed by adding dummy variables to the equation representing residence in each city. We find that significant effects do indeed remain, above the .001 level for all preference items, even if the net increment in explained variance ranges only from 2.0 to 6.2%.

One question from our initial set of hypotheses remained. Does the relative contribution of explanatory variables interact with the degree of separableness as suggested in Proposition G? To examine this proposition further, regressions for each of four policy types were computed in step-wise fashion to decompose the variance attributable to the three sets of explanatory variables. The top of Table 8 shows the adjusted R^2 associated with individual, neighborhood, and city variables and shared variances. The variance explained by all three sets simultaneously included in the equation are shown in row H. At the bottom of the table are results for testing Proposition 6. We calculated the ratio of the variance of individual variables alone and that of city variables alone to the maximum variance explained. The relative predictive power of

TABLE 8: ANALYSIS OF VARIANCE FOR INDIVIDUAL, NEIGHBORHOOD, AND CITY-LEVEL VARIABLES[a]

	Environment Services	Traditional Services	Public Schools	Welfare Services
a. Percentage of Variance Explained				
A. Individual variables alone	2.8	6.3	8.7	15.3
B. Neighborhood variables alone	2.7	1.9	.2	6.2
C. Neighborhood addition over A	1.3	.4	.2	.6
D. Shared variance of A and B	1.5	1.5	0	5.6
E. City alone (dummy variables)	3.7	6.2	3.9	4.6
F. City addition to A, B	2.2	4.5	2.6	1.8
G. Shared variance of city and A, B	1.5	-1.7	1.3	2.8
H. Total-individual, neighborhood and city	6.3	11.2	11.5	17.7
b. Proportion (percentage) of Total Variance Explained Attributable to Individual and City Variables				
1. Ratio A/H	44	56	76	86
2. Ratio E/H	59	55	39	26

[a]The variables are the same as those presented in Table 7 except "perceived distribution variables" were not included in the regression. Thus total R^2s in row H are slightly different than those shown in Table 7.

individual variables, as shown in row 1 increases for more separable goods. The relative predictive power of city (row 2) shows the opposite results, indicating that the less separable the policy, the more likely that variables associated with cities (rather than neighborhoods or individuals) explain variation in citizen preferences. This finding lends support to our classification scheme of policies along a separable continuum.

CONCLUSION

Citizen demands for public goods like environmental services are affected less by changes in population characteristics of individual citizens than by the actual policy provided in the city. By contrast, as population changes occur, demands for separable goods like welfare are likely to change more in keeping with the population change. Survey analysts interpreting their results, particularly those who seeks to generalize from national surveys to the local level, or to project current or future attitudes from past surveys in the same city, would do well to keep these differential patterns in mind. These results also suggest the importance of a public-separable good continuum for analyses of urban policy outputs (Clark, 1973), as in the proposition: centralization encourages public goods.

NOTES

1. Note that an individual is "affluent" here if his property value is above the mean for his city; and an individual below the mean is "poor." Such contextual effects whereby the affluent subsidize other citizens will not emerge if cities are homogeneous in terms of their citizens' taxable wealth. Similarly, the less homogeneous the city in terms of citizens' taxable wealth, the greater should be the subsidization effects, and the greater the political conflict over distribution of services—ceteris paribus, of course. This last proposition may partially explain the apparently paradoxal results that voting turnout is *negatively* related to the average educational level of citizens across the 282 cities studied by Alford and Lee (1968), since many of these cities with highly educated citizens are presumably semihomogeneous suburbs.

2. Proposition C depends not on the technical aspects of producing public or separable goods, but the more political issue of how to allocate separable goods. Hypothetically, a city government could allocate separable goods in proportion to taxes paid. Moving in this direction as far as competing cities should reduce incentives which the affluent have to migrate to suburbs or sunbelt (Buchanan, 1971). Empirically, however, leaders in most American cities seem to feel more constrained to allocate separable goods equally across income groups, and in the case of the separable goods studied here, to allocate them disproportionately to less affluent citizens. More affluent citizens are thus likely to benefit more from, and favor more spending on, public than separable goods in most American cities. Hence Proposition C.

3. We are grateful to Lawrence A. Williams of the Urban Observatory Program of the National League of Cities/U.S. Conference of Mayors for making the study data available. The ten cities were: Albuquerque, Atlanta, Baltimore, Boston, Denver, Kansas City, KS, Kansas City, MO, Milwaukee, Nashville, San Diego.

4. The three response categories of "spend more," "about as much," and "spend less" were scored 1, 0, and −1 respectively for use in further analyses. No answer and don't knows were not analyzed. Four other items included in the list of services were not analyzed as they could not be linked to specific policy

outputs. Measurement problems for citizen preferences using items such as these have been discussed elsewhere (cf., Clark, 1973a, 1974, 1976).

5. Here and below the policy issue measures are z-scores of the factor patterns for "environmental," "traditional," and "welfare" services as shown in Table 6, and z-scores of responses to the questionnaire items about public schools. Overall spending was the z-score version of the first principal component of the orthogonal factor analysis of the ten policy items.

6. Elsewhere we have analyzed these and related data in considerably greater detail in search of curvilinear relationships. For example, all ten policy issues were analyzed separately; residuals from regressions controlling certain individual and neighborhood characteristics were plotted against income. Separate plots were drawn for homeowners and renters, for citizens consuming the service and nonconsuming citizens, etc. Policy responses were also adjusted in various ways, such as by subtracting "spend more" from "spend Less" responses. See Hoffman (1975, 1976) for most of these results. Virtually no support for the curivlinear hypotheses was uncovered in these efforts. James Vanecko and Peter Lupsha (in unpublished papers) have also analyzed the Urban Observatory data and generally found minimal curvilinear relationships between spending preferences and income. Empirical estimates for the separate coefficients of 1 have not been attempted because of the low number of policy issues.

7. This is similar to results of other studies of urban citizen policy preferences: Schuman and Gruenberg (1970) explained an average of 9% of the variance in 26 items; Rossi, Berk, and Eidson (1974) explained about 15% of the variance in numerous items; McIver and Ostrom (1976) explained less than 10% of the variance in several items on the police; Curry (1976) explained an average of 6% of the variance for eight items from Wilson and Banfield's (1971) Boston survey. Nevertheless, many relationships in all these studies were strongly significant using standard criteria of statistical inference.

8. Further analysis of Blacks' preferences for spending, which considers income, neighborhood, and strategic behavior effects is in Hoffman (1975, 1976).

9. Irish Catholics in particular did not favor more spending than most other respondents; the explanation for high spending by cities with numerous Irish Catholic residents thus seems not to be a consequence of citizen policy preferences, but related to more general dimensions of political culture, including a preference for separable over public goods (Clark 1975).

REFERENCES

AARON, H. J. (1975) Who Pays the Property Tax? Washington, DC: Brookings Institution.
ALFORD, R. R. and E. C. LEE (1968) "Voting turnout in American cities." Amer. Pol. Sci. Rev. 62: 796-813. .
BANFIELD, E. C. and J. Q. WILSON (1963) City Politics. Cambridge, MA: Harvard University Press.

BERGSTROM, T. C. and R. P. GOODMAN (1973) "Private demands for public goods." Amer. Econ. Rev. 63: 280-296.
BRADFORD, D. F. (1971) "Joint products, collective goods, and external effects: comment." J. of Pol. Economy 79: 1119-1128.
BUCHANAN, J. M. (1971) "Principles of urban fiscal strategy." Public Choice 11: 1-17.
CLARK, T. N. [ed.] (1976) Citizen Preferences and Urban Public Policy. Beverly Hills: Sage Publications.
--- (1975) "The Irish ethic and spirit of patronage." Ethnicity 2: 305-359.
--- (1974) "Can you cut a budget pie?" Policy and Politics 3: 3-32.
--- (1973) Community Power and Policy Outputs. Beverly Hills: Sage Publications.
--- (1973a) "Community social indicators: from analytical models to policy applications." Urban Affairs Q. 9: 3-36.
CONVERSE, P. (1964) "The nature of belief systems in mass publics," in D. Apter (ed.) Ideology and Discontent. New York: Free Press.
CURRY, G. D. (1976) "Utility and collectivity: some suggestions on the anatomy of citizen preferences," in T. N. Clark (ed.) Citizen Preferences and Urban Public Policy: Models, Measures, Uses. Beverly Hills: Sage Publications.
DAVIS, O. A. and G. H. HAINES (1966) "A political approach to a theory of public expenditure: the case of municipalities." National Tax J. 19: 259-273.
FOWLER, F. J., Jr. (1974) Citizen Attitudes Toward Local Government, Services, and Taxes. Cambridge, MA: Ballinger.
HOFFMAN, W. L. (1976) "The democratic response of urban governments: an empirical test with simple spatial models," in T. N. Clark (ed.) Citizen Preferences and Urban Public Policy: Models, Measures, Uses. Beverly Hills: Sage Publications.
--- (1975) Citizen Policy Preferences and Urban Government Performance. Ph.D. dissertation, Department of Political Science, University of Chicago.
McIVER, J. P. and E. OSTROM (1976) "Using budget pies to reveal preferences: validation of a survey instrument," in T. N. Clark (ed.) Citizen Preferences and Urban Public Policy: Models, Measures, Uses. Beverly Hills: Sage Publications.
NIE, N. H., S. VERBA, and J. PETROCIK (1976) The Changing American Voter. Cambridge, MA: Harvard University Press.
PAGE, B. I. (1977) "Elections and social choice: the state of the evidence." Amer. J. of Pol. Sci. 21: 639-662.
PHILLIPS, K. B. (1970) The Emerging Republican Majority. New York: Doubleday-Anchor.
ROSSI, P. H., R. A. BERK, and B. K. EIDSON (1974) The Roots of Urban Discontent. New York: John Wiley.
SAMUELSON, P. A. (1954) "The pure theory of public expenditure." Rev. of Economics and Statistics 36: 387-89.
SCHUMAN, H. and B. GRUENBERG (1970) "The impact of city on racial attitudes." Amer. J. of Sociology 76: 213-261.
STIGLER, G. J. (1970) "Director's law of public income redistribution." J. of Law and Economics 13: 1-10.
WEICHER, J. C. (1971) "The allocation of police protection by income class." Urban Studies 8: 207-220.

WILSON, J. Q. (1975) "The riddle of the middle class." Public Interest 39: 125-129.
——— and E. C. BANFIELD (1971) "Political ethos revisited." Amer. Pol. Sci. Rev. 65: 1048-1062.
——— (1964) "Public-regardingness as a value premise in voting behavior." Amer. Pol. Sci. Rev. 58: 876-887.

6

The Season of Tax Revolt

JOHN L. MIKESELL

□ PUBLIC INFLUENCE ON STATE AND LOCAL FISCAL decisions has historically been through the selection of representatives who made budget decisions, revenue choices, and stood for reelection on the basis of that performance. Because most state and local governments borrow only for capital expenditures—and often only under rigid ceilings or voter approval requirements—with operating budgets kept in balance, these representatives could not easily unleash themselves from citizen scrutiny: more money spent meant obvious tax rate increases with potential voter displeasure.[1] The late 1970s have, however, produced a public attitude that the traditional control method is insufficient and that strong absolute limits are required. Regardless of whether the representative model has actually failed, the several tax or expenditure control referenda of 1978—the most recent reflection of this attitude—will alter urban government fiscal process for many years.

As it turns out, the citizen reaction, while frequently harsh and poorly structured, is not without logical motivation, at least in the short run. The historical focus of revenue-raising control—and hence on

AUTHOR'S NOTE: *James Skubic gathered all referendum information from individual states.*

spending power—has been on statutory tax rates and rate structures.[2] This focus on rates may be of minimal consequence for control of public activity in times of rapidly expanding tax bases. If a state applies an income tax structure with graduated rates and fixed dollar exemptions, that government can garner even greater percentages of the tax base as inflation drives incomes higher in dollar, but not necessarily real, amount.[3] In a similar way, property tax rate limits—at least forty-one states had local units operating under legal limits in 1976 (ACIR, 1977a: 151-167)—may lose their power when an assessment system is able to capture inflation-fueled market value increases or when reassessment brings a dramatic single-year increase in property assessments (Mikesell, 1978). In this environment, limits on overall spending, special controls on effective (not statutory) property tax rates, or tax levies may seem the only public alternative to the higher tax collections and higher spending which the traditional budget process does not restrain. We see, for example, the property tax rate freeze for California counties, cities, and special districts of 1972 rendered nearly meaningless by rapid property assessment growths with the absolute controls of Proposition 13 the apparent solution. The representative process may eventually produce rate adjustments to correct for tax base inflation, but the fiscal mood of the public has been to take no chances on that uncertain outcome.

The voter-driven control measures of 1978, while unique in overall success, did not emerge without precursors. A few states, Michigan and California the most famous examples,[4] had previously held statewide referenda on spending limit proposals which were decisively defeated. More importantly, state legislatures had placed legal spending limits on themselves in three states. In 1977, Michigan established a budget stabilization fund to restrict state expenditure to income growth, and Colorado constrained itself to an annual 7% increase in general fund spending with an additional 4% to a reserve fund (amounts over 11% refunded to taxpayers). In 1976, New Jersey restrained its spending to an increase no greater than the increase in state personal income. And in 1978, Colorado, possibly in response to the potential need for continuing refunds, indexed its state personal income tax to prevent inflation from pushing taxpayers into higher brackets (Shannon and Weissert, 1978). Similar limits on tax collections or expenditure were also in place among local governments. At least fourteen states and the District of Columbia have enacted laws limiting the property tax levy or total spending by one or more local units of government since 1970

(ACIR, 1977b: 2). Beyond these, at least four states—Florida, Montana, Virginia, and Ohio—have "full disclosure" elements in their property assessment law to force rate reductions when mass revaluation produces dramatic assessment increases.

THE CONTROL REFERENDA

Possibly more interesting, however, is the wave of state referenda presented in 1978 that limit or would have limited the aggregate spending or taxing authority of state and/or local government. While the California Proposition 13 initiative received greatest popular attention, sixteen other states held such votes in the year. These actions—beginning in March (Tennessee) and ending in November (fifteen states)—took several different forms during the season of revolt, as Table 9 identifies and summarizes and Table 10 classifies. The measures are often jumbled, covering an assortment of tax or spending controls, but a tentative classification can be proposed. First, two states—Massachusetts and Illinois—voted on advisory petitions with no necessary consequences of the outcome. The Massachusetts referendum, which took slightly different forms in different senatorial districts, inquired about attitudes toward limits on state and local spending as a guide to decisions to be made by individual legislators. Not all districts voted on the question, but around 95% of the state's population was covered.[5] The Illinois question asked whether there should be limits on taxes and spending applied to state, local, and school district governments. Action is not guaranteed from either result, but it is reasonable to expect some legislative action. The Massachusetts legislature did, in fact, pass a tax limitation amendment in 1978 which, if approved again during the 1979 session, will appear on the 1980 ballot (Tax Administrators News, 1978: 110).

Second, eight states—Alabama, California, Idaho, Massachusetts, Michigan, Missouri, Nevada, and Oregon—held referenda dealing primarily with property tax limitation. All states but Idaho framed constitutional amendments, and the Idaho attorney general has held that their revision violates uniformity clauses of the state constitution.[6] California (Proposition 13), Measure J in Michigan, Nevada, and Measure 6 in Oregon all limit the property tax rate and define the manner in which assessed values can grow over time (often with initial reductions to a prior year level), thus defining a ceiling property tax budget. Alabama

TABLE 9: 1978 REFERENDA ON STATE OR LOCAL TAXING OR SPENDING LIMITATIONS

State	Measure Name (Month of Vote)	Source (I=initiative; L=legislative; C=constitutional correction)	Short Description	Result (P=pass; F=fail)
Alabama	Act 6 (HB170), second special session (November)	L	Constitutional amendment to alter property classification scheme, lower assessment ratios, set limits on effective property tax rates, and create additional property tax exemptions	P:60% for
Arizona	S.C.R. 1002 (November)	L	Constitutional amendment to limit state appropriations to 7% state personal income	P:78% for
California	Proposition 8 (June)	L	Constitutional amendment to permit owner-occupied dwellings to be taxed at lower rate than that levied on other property	F:53% against
California	Proposition 13 (June)	I	Constitutional amendment to limit real property taxes to 1% of value, to restrict the real assessed value base, to require popular votes on new local taxes, and to require 2/3 legislative vote on state tax increases	P:64% for
Colorado	Amendment no. 2 (November)	I	Constitutional amendment to limit annual increase in per capita state and local expenditures to percentage	F:58% against

110

TABLE 9 (cont.)

State	Measure Name (Month of Vote)	Source (I=initiative; L=legislative; C=constitutional correction)	Short Description	Result (P=pass; F=fail)
			increase of cost of living, unless voters authorize otherwise, and limit excess of state receipts over expenditures to 5% of aggregate appropriations	
Hawaii	Amendment no. 11 (November)	C	Constitutional amendment to require legislature to establish ceiling limiting rate of growth of general fund appropriation (excluding federal funds) to growth of state economy with exceptions only by 2/3 legislative vote and to require tax refund or credit when general fund balance at close of 2 successive fiscal years exceeds 5% of general fund revenues for those years.	P:67% for
Idaho	Initiative petition no. 1 (November)	I	Proposal to limit property tax, to 1% of "actual market value," defined as 1978 market value or appraisal value on later sale, new construction, or change in ownership (statutory)	P:58% for

TABLE 9 (cont.)

State	Measure Name (Month of Vote)	Source (I=initiative; L=legislative; C=constitutional correction)	Short Description	Result (P=pass; F=fail)
Illinois	Advisory poll (Novmeber)	I	Petition asking whether legislature should act and consitution be amended to impose ceiling on taxes and spending by state, local governments, and school districts (advisory only)	P:82% for
Massachusetts	H.B. 5317 (November)	L	Constitutional amendment to permit classification of property according to use for tax purposes	P:66% for
Massachusetts	Conway-Carlson (November)	L	Advisory petition asking whether voters favor limits on state and local spending (on ballot in about 95% of state); nonbinding public policy question	P:79% for
Michigan	Headlee Amendment (Measure E) (November)	I	Constitutional amendment to limit state taxes to present ratio to state personal income, with change by 2/3 vote of electorate or legislative override in declared emergency	P:52% for
Michigan	Tisch Amendment (Measure J) (November)	I	Constitutional amendment to limit assessment ratio to 25%, limit state income tax rates to 5.6% (presently	F:63% against

TABLE 9 (cont.)

State	Measure Name (Month of Vote)	Source (I=initiative; L=legislative; C=constitutional correction)	Short Description	Result (P=pass; F=fail)
Missouri	S.J.R. 37 (November)	L	at 4.6%), and allow school district income taxes up to 1% on voter approval Constitutional amendment to permit general assembly to require political subdivisions to reduce their property taxes	P:67% for
Nebraska	Initiated measure no. 302 (November)	I	Constitutional amendment to limit local budget increases to 5%, with limited exceptions	F:56% against
Nevada	Question 6 (November)	I	Constitutional amendment to limit property tax to 1% of "full cash value" (1975) or appraisal value on later sale, construction, or ownership change (must be resubmitted to voters in November 1980 for final approval of amendment)	P:78% for
North Dakota	Initiative measure no. 2 (November)	I	Intiative decreasing personal income tax rates and increasing corporate tax rates (statutory)	P:65% for
Oregon	Measure 6 (November)	I	Constitutional amendment to limit property taxes to 1 1/2% of "full	F:52% against

TABLE 9 (cont.)

State	Measure Name (Month of Vote)	Source (I=initiative; L=legislative; C=constitutional correction)	Short Description	Result (P=pass; F=fail)
Oregon	Measure 11 (H.J.R. 84, special session) (November)	L	cash value" (1975) or appraisal value on later sale or new construction. State payment of 1/2 property tax on owner-occupied principal residences (up to $1500) with equivalent renter relief, limitation on state and local spending, requirement of 2/3 legislative vote on tax increases, and real property assessment freeze for one year (constitutional amendment)	F: 55% against
South Dakota	Proposition D (November)	L	Constitutional amendment to require voter consent ot 2/3 legislative vote to increase rate of tax on personal income, corporate income, or sales and services, or maximum allowable levies or valuation basis for real or personal property	P: 53% for
Tennessee	Proposal 9 (March)	L	Constitutional amendment prohibiting growth rate of appropriations from tax revenues from exceeding estimated rate of growth in state	P: 65% for

TABLE 9 (cont.)

State	Measure Name (Month of Vote)	Source (I=initiative; L=legislative; C=constitutional correction)	Short Description	Result (P=pass; F=fail)
			economy with majority vote of legislative override.	
Texas	H.J.R.1, first special session (November)	L	Constitutional amendment to provide property tax relief, ceilings on local property taxes, and a limit on state spending	P:84% for

NOTE: Excludes New Mexico property tax relief for elderly and disabled (legislative referendum rejected), Virginia relief for renovated or replaced real estate (legislative referendum passed), West Virginia exemption for inventory and warehoused goods (legislative referendum rejected), Arkansas elimination of sales tax on food and medicine (initiative rejected), California prevention of reassessment of property reconstructed after a disaster (legislative referendum passed), and Florida revision of finance and taxation section of Constitution to expand property tax relief and make other revisions (constitutional convention proposal defeated).

SOURCE: State Tax Review (many 1978 issues), copies of the referenda, and correspondence or conversation with state officials.

achieved a similar effect through reduced assessment ratios and limits on effective property tax rates. The Missouri amendment makes it possible for the state legislature to reduce local property taxes, regardless of prior authorization for the rate in Constitution or statute. The Alabama and Missouri actions apparently were both induced by concerns about the property tax impact of general reassessments in the state. Finally, Massachusetts authorized the state legislature to adopt property tax classification according to use, apparently to generate lower residential and higher business taxes.

Third, eight states—Arizona, Colorado, Hawaii, Michigan, Nebraska, Oregon, Tennessee, and Texas—voted on referenda which were primarily spending limits. The measures frequently were designed to restrict spending totals to a specified fraction of state personal income, although the Colorado limit was tied to population and the cost of living, the Tennessee limit was related to the estimated growth in the state economy, and Nebraska set the limit at 5% per year. Arizona, Hawaii,

TABLE 10: CLASSIFICATION OF STATES BY REFERENDA TYPE, 1978

Primarily Property Tax Limit	Global Advising Poll
Alabama	Illinois
California	Massachusetts
Idaho	
Massachusetts	
Michigan	
Missouri	
Nevada	
Oregon	

Primarily Spending Limit	Primarily Rate Revision
Arizona: state	North Dakota
Colorado: state, local	
Hawaii: state	
Michigan: state	
Nebraska: local	
Oregon: state, local	
Tennessee: state	
Texas: state, local	

Primarily Method for Rate Increase
South Dakota

SOURCE: Compiled from Table 9.

Tennessee, and Measure E in Michigan were strictly state limits. The Nebraska limit was strictly local. The remaining three states considered limits on both state and local spending.

Two other states held significant limitation or revision referenda that do not fall neatly into the other classes. South Dakota considered an amendment to require referenda approval or an extraordinary majority (two-thirds) in the legislature to increase the rate of tax on personal or corporate income, sales or services, or real or personal property and the valuation basis for property. And North Dakota voted on reduced personal income tax rates with increased corporate income tax rates. While none of the measures match the classic lines of tax revolt, it would be inappropriate to exclude them from the listing.

OBSERVATIONS ON REFERENDA

A careful analysis of the referenda results from 1978 may provide insights into public behavior which may suggest trends and useful government responses. This section examines selected characteristics of states holding referenda to attempt to understand the underlying forces. Particular elements examined are the relationship between basic government structure and referenda, the success record, the fiscal environment of referenda states, and patterns of regional voting. Evidence from these considerations can suggest meaningful public responses in anticipation and reaction to voter rebellion. It is reasonable to expect as a working hypothesis that states acting to control particular fiscal variables will have been experiencing levels of these variables greater than the nation as a whole—if there is any rational basis to the revolt movement.

REFERENDA AND GOVERNMENT STRUCTURE

The referenda reached the 1978 ballots as a result of legislative action, constitutional approval, and petition by voters. Lumping all proposals for the year, eleven emerged from convention or legislature and eleven from popular petition. Four proposals by popular petition were defeated; three from representative bodies failed. Overall, popular initiatives were presented in ten states, with six involving constitutional amendment. Referenda from representative bodies were presented in eleven different states, with ten involving constitutional amendment. As only seventeen state constitutions authorized constitution amendment by initiative (Strum, 1978: 210), the record of six states voting during the same year on the rebellion issue is impressive. Even more notable is the success record of these initiatives: "Often initiative proposals that originate by popular petition lack the substantial popular support necessary to assure their success. Thus, the rate of adoption is usually substantially lower than that of legislative proposals" (Strum, 1978: 195). In the period from 1968 through 1977, almost 26% of the fifty-eight constitutional initiatives on all topics were approved; the 1978 success rate for constitutional initiatives on fiscal control was 43%. All such initiatives—constitutional, legislative, and adivsory—had a successful rate of 60%. The legislative proposal amendment success rate of 77% is approximately that of the entire 1968 through 1977 period, but since two legislative proposals (defeated in Oregon and California) were placed on the ballot as substitutes for voter initiated items, conclusions are difficult.

FISCAL ENVIRONMENTS

The fiscal environments of states differ dramatically, as influenced by both taste for public services and prevailing affluence. Table 11 presents levels and state ranks for effective property tax rates on single family houses, property, and total state-local tax burdens on income, per capita, state, and state-local direct general expenditure, and state and state-local direct general spending increase. These seem reasonable measures to compare because they are the general fiscal measures which the 1978 referenda sought to control. The ranks of the states holding referenda are separately displayed in Table 12, along with the number of states in the group falling above the median in the nation, to indicate the forces that may have induced the referenda.

An apparent observation is that the spending limit referenda were not held uniformly in high-spending or dramatic spending increase states, and the property tax limit referenda were not in the highest property tax states. For instance, property tax-controlling referenda were held in the state with the lowest property tax burden (Alabama) and in the second highest (Massachusetts). Similar dispersion—although not quite so dramatic—appears with spending limitation. There are, however, two characteristics which show a consistent and reasonable pattern in the affected states. First, eight of the ten property tax-limiting states have effective property tax rates on single family homes

TABLE 11: TAXES AND STATE SPENDING BY INDIVIDUAL STATES

State	Effective Property Tax Rate 1975[a]		Property Taxes per $1000 Personal Income 1976-1977[b]		State-Local Taxes per $1000 Personal Income 1976-1977[b]	
	Percentage	Rank	Dollars	Rank	Dollars	Rank
Alabama	0.75	45	11.76	50	99.96	49
Alaska	1.73	20	134.81	1	234.83	1
Arizona	1.54	27	55.29	14	144.14	8
Arkansas	1.41	31	22.56	44	101.78	48
California	2.08	11	65.14	4	154.93	3
Colorado	1.99	14	49.47	16	129.72	13
Connecticut	1.94	15	55.92	13	119.97	26
Delaware	0.92	43	19.06	48	117.96	32
Florida	1.18	39	35.19	33	104.74	46
Georgia	1.33	32	34.70	34	111.50	37
Hawaii	NA	—	24.07	42	140.71	10
Idaho	1.86	16	37.45	31	116.97	33
Illinois	2.21	7	45.09	21	118.99	28
Indiana	1.64	22	39.21	25	105.41	45
Iowa	2.20	8	46.71	19	120.25	25
Kansas	1.55	26	46.57	20	113.24	35
Kentucky	1.23	37	21.11	46	112.76	36
Louisiana	0.64	46	18.73	49	120.91	24
Maine	1.86	16	44.94	22	124.39	21
Maryland	2.01	13	38.54	28	129.47	14
Massachusetts	3.26	1	74.24	2	151.36	6
Michigan	2.38	6	49.28	18	130.39	12
Minnesota	1.58	24	43.94	23	146.92	7
Mississippi	1.12	40	26.03	39	118.17	31
Missouri	1.85	19	32.54	35	120.60	47

TABLE 11 (cont.)

State	Effective Property Tax Rate 1975[a]		Property Taxes per $1000 Personal Income 1976-1977[b]		State-Local Taxes per $1000 Personal Income 1976-1977[b]	
	Percentage	Rank	Dollars	Rank	Dollars	Rank
Montana	1.60	23	64.30	5	136.05	11
Nebraska	2.50	5	58.94	11	127.84	17
Nevada	1.53	28	41.39	24	129.30	15
New Hampshire	NA	—	65.70	3	106.23	43
New Jersey	3.15	2	63.36	6	126.06	19
New Mexico	1.56	25	21.71	45	119.54	27
New York	2.56	4	63.30	7	176.84	2
North Carolina	1.51	30	25.88	40	109.83	38
North Dakota	1.53	28	38.82	26	118.35	30
Ohio	1.29	35	38.34	29	99.44	50
Oklahoma	1.27	36	23.99	42	106.53	42
Oregon	2.18	9	57.53	12	129.25	16
Pennsylvania	1.71	21	31.03	37	118.80	29
Rhode Island	NA	—	52.06	15	126.37	18
South Carolina	1.07	42	25.17	41	107.67	40
South Dakota	2.14	10	60.19	10	123.46	22
Tennessee	1.31	34	26.89	41	107.27	41
Texas	2.06	12	38.55	27	105.61	44
Utah	1.20	38	36.79	32	125.88	20
Vermont	NA	—	61.97	9	151.84	5
Virginia	1.32	33	31.26	36	108.69	39
Washington	1.86	16	38.04	30	122.34	23
West Virginia	0.78	44	20.98	47	116.39	34
Wisconsin	2.63	3	49.29	17	143.61	9
Wyoming	1.12	40	62.98	8	154.76	4

120

TABLE 11 (cont.)

State	State Direct General Expenditure Per Capita 1976-1977[c]		State-Local Direct General Expenditure Per Capita 1976-1977[c]		State Spending Increase, 1966-1967 to 1976-1977[c,d]		State-Local Spending Increase, 1966-1967	
	Dollars	Rank	Dollars	Rank	Percentage	Rank	Percentage	Rank
Alabama	504.60	29	1006.70	44	195.8	23	190.2	23
Alaska	1950.40	1	3275.40	1	263.6	7	309.2	1
Arizona	420.00	43	1242.80	22	152.2	42	234.2	7
Arkansas	441.90	40	876.20	50	173.0	36	181.7	33
California	454.10	38	1485.80	5	148.9	45	160.9	43
Colorado	484.40	32	1345.80	15	214.3	15	230.8	9
Connecticut	517.10	25	1151.60	31	184.7	30	159.0	44
Delaware	815.10	3	1458.20	8	181.0	34	167.5	40
Florida	341.50	50	1098.90	37	227.0	13	256.0	2
Georgia	420.20	44	1002.60	43	207.5	20	199.3	19
Hawaii	1467.60	2	1915.20	2	259.6	5	247.9	4
Idaho	541.50	20	1140.80	32	211.7	16	209.0	15
Illinois	507.20	28	1266.10	21	259.3	6	213.0	13
Indiana	347.40	49	953.10	48	150.3	44	146.7	47
Iowa	483.40	33	1235.30	23	158.6	40	170.6	37
Kansas	491.40	31	1193.30	26	229.6	11	177.0	36
Kentucky	574.40	16	1006.20	41	180.3	35	165.9	42
Louisiana	585.10	15	1207.20	24	150.8	43	166.8	41
Maine	533.10	22	1119.70	33	182.0	33	201.7	17
Maryland	510.50	26	1452.80	9	299.2	3	245.2	5
Massachusetts	587.10	14	1378.10	13	348.3	2	215.3	11

TABLE 11 (cont.)

State	State Direct General Expenditure Per Capita 1976-1977[c]		State-Local Direct General Expenditure Per Capita 1976-1977[c]		State Spending Increase, 1966-1967 to 1976-1977[c,d]		State-Local Spending Increase, 1966-1967	
	Dollars	Rank	Dollars	Rank	Percentage	Rank	Percentage	Rank
Michigan	525.90	23	1389.80	12	209.2	17	189.9	24
Minnesota	509.10	27	1459.90	7	235.0	10	196.4	22
Mississippi	475.30	35	1017.80	40	209.0	18	198.9	21
Missouri	362.60	48	942.20	49	143.0	47	143.2	49
Montana	654.90	7	1409.10	11	182.7	31	200.1	18
Nebraska	435.70	41	1152.50	30	193.9	26	185.6	29
Nevada	558.30	17	1469.80	6	187.8	28	199.3	20
New Hampshire	535.80	21	1116.50	29	236.2	7	240.0	6
New Jersey	467.20	36	1326.60	16	352.2	1	233.1	8
New Mexico	557.60	18	1176.70	28	133.7	48	149.1	46
New York	456.10	37	1795.20	3	194.9	25	185.5	30
North Carolina	388.80	46	982.10	46	208.6	19	214.7	12
North Dakota	712.90	5	1308.30	18	144.5	46	129.0	50
Ohio	368.50	47	1109.40	35	199.1	21	188.8	25
Oklahoma	498.20	30	1044.80	39	125.3	49	152.8	45
Oregon	587.90	12	1414.00	10	195.2	24	212.4	14
Pennsylvania	524.50	24	1166.40	34	227.5	12	187.4	28
Rhode Island	686.10	6	1283.10	19	165.1	38	169.4	39
South Carolina	542.30	19	978.50	47	296.4	4	255.0	3
South Dakota	623.50	8	1180.40	27	157.9	41	144.7	48

TABLE 11 (cont.)

State	State Direct General Expenditure Per Capita 1976-1977[c]		State-Local Direct General Expenditure Per Capita 1976-1977[c]		State Spending Increase, 1966-1967 to 1976-1977[c,d]		State-Local Spending Increase, 1966-1967	
	Dollars	Rank	Dollars	Rank	Percentage	Rank	Percentage	Rank
Tennessee	426.40	42	992.30	45	196.3	22	183.9	31
Texas	400.10	45	1003.40	42	224.2	14	207.4	16
Utah	592.20	11	1201.10	25	182.6	32	187.7	27
Vermont	775.40	4	1279.50	20	162.8	39	181.5	34
Virginia	478.30	34	1104.50	36	236.3	8	229.4	10
Washington	619.90	9	1357.00	14	190.9	27	188.6	26
West Virginia	607.40	10	1083.20	38	166.8	37	179.6	36
Wisconsin	449.10	40	1321.70	17	184.8	29	182.2	32
Wyoming	587.90	12	1572.20	4	95.7	50	170.6	38

SOURCES: [a]Advisory Commission on Intergovernmental Relations (1977b:107)
[b]National Journal (1978; 1973).
[c]U.S. Bureau of the Census (1978).
[d]U.S. Bureau of the Census (1968).

greater than the national median. It is reasonable to expect that this condition would stimulate organized efforts to control or even reduce property tax levels. One should also note that, of the other fifteen states above the median on this measure (data are unavailable for four states), three held referenda on broad spending limitations and of the other states, eleven have no provision for constitutional amendment by initiative (Smolka, 1978: 243), and seven have no provision for legislation by initiative (Strum, 1978: 209, 210). The pattern is not as strong

TABLE 12: RANKS OF 1978 LIMIT REFERENDA STATES ON CONTROL VARIABLES

Property Tax Limit	Effective Rate	Property Tax Burden	Total Tax Burden
Alabama	45	50	49
California	11	4	3
Idaho	16	31	33
Massachusetts	1	2	6
Michigan	6	18	12
Missouri	19	35	47
Nevada	28	24	15
Oregon	9	12	16
(Illinois)	7	21	28
(South Dakota)	10	10	22
Number above median	8	7	6

Spending Limit	State General Expenditure/ Capita	Increase in State General Expenditure	State-Local General Expenditure/ Capita	Increase in State-Local General Expenditure
Arizona (S)	43	42	22	7
Colorado (S, L)	32	15	15	9
Hawaii (S)	2	5	2	4
Michigan (S)	23	17	12	24
Nebraska (L)	41	26	30	29
Oregon (S, L)	12	24	10	14
Tennessee (S)	42	22	45	31
Texas (S, L)	45	14	42	16
(Illinois)	28	6	21	13
(Massachusetts)	14	2	13	11
Number above median	4	8	7	8

SOURCE: Computed from Table 11.

for total tax or property tax collections per dollar of state personal income. As much of the total tax may be initially paid by business (and not apparently paid by individual voters) or designed to be exported to nonresidents, it is not unreasonable for this absolute burden influence to be milder. The effective property tax rates are high in the spending limit states as well: overall, seven of the nine states for which there are data are above the national median. Including states voting only on spending limits—no inclusion in the property tax limit group—three of five are above the median. Evidence suggests that voters are upset with high property tax rates on homes and translate this irritation into referenda where possible.

Second, increases in state and local spending appear to have an impact on the incidence of spending limit referenda. As Table 12 shows, eight of the states with such referenda have above median state spending growth from 1966 to 1976. No similar strength of pattern appears for per capita levels. Of the seventeen remaining states above the median, two (Alabama and Idaho) held a property tax limit referendum, thirteen have no provision for constitutional amendment by initiative, and twelve have no provision for referenda on legislation by initiative. Growth in state spending is not, however, closely associated with property tax-limiting referenda: Only six of the ten states in that group showed spending growth above the median, and only two of that six were not also facing a spending referendum. In sum, spending growth did not appear to be a serious factor in the property tax-limiting referenda states, while effective property tax rates cannot be dismissed as an influence in any of the referenda states.[7]

GEOGRAPHIC PATTERNS

One striking feature of the 1978 referenda is the definite regional patterns shown by the states holding them. One state (Massachusetts) among the seventeen from Maine to Florida (the New England, Middle Atlantic, and South Atlantic states, by census classification) held a statewide referendum. New Jersey does have budget limitations from 1976 legislation, as previously described. Among the twenty states in the central area (East South, East North, West North, and West South Central), nine held limit referenda. In the Pacific and Mountain areas, seven states of the thirteen held referenda. Overall, the referenda pattern appears heavily among the central and western states; it is not yet a phenomenon of the eastern United States.

As a final note on the referenda patterns, it may be useful to consider the urbanization patterns of the states holding the 1978 referenda. In sum, eight states had percentages of population urban (1970) below the national level and nine had higher percentages.[8] There are similarly no significantly different patterns when the two major varieties of limits are compared separately. Generally, urban or nonurban states seem equally susceptible to limitation processes. Of course, large, urban state adoption has impacts on far more individuals than does rural state limitation, but there appears to be no special difference in the likelihood of adoption of control.

PROSPECTS

Despite the six failed proposals, the forces of revolt did not miss total victory in 1978 by much. In most instances, the defeated proposal was less stringent—or appeared to be so—than other options or existing conditions. The defeated Proposition 8 in California was milder than the successful Proposition 13. Amendment No. 2 in Colorado would have tied per capita spending increases to the cost of living; legislation from 1976 and 1977 restrained property tax levies (in most units) and state spending to a 7% increase. The Tisch Amendment (Measure J) in Michigan failed, but the Headlee Amendment (Measure H) passed. Furthermore, Measure J explicitly permitted an increase in the state income tax from 4.6% to 5.6% and allowed school districts to levy new income taxes of up to 1% on voter approval. The tax revolt attitude to Measure J thus is somewhat clouded. Nebraska rejected Measure No. 302—to limit local budget increases to 5%—but 1977 legislation had already limited increases to 7%. The remaining failure of tax limits was in Oregon, where both moderate and stringent limitations were rejected. In sum, the prevailing voter mood was for limitations and the group of referenda failures represent no significant counter evidence. More restraint seems likely, limits—constitutional or statutory—are probably going to appear in one form or another in most states, and, in response to pressures of inflation, limits are going to deal with total tax paid or total dollars spent, not just on statutory rates.

Some forces for future control are already in place. A Federation of Tax Administrators survey prior to the November election located a number of states which, while not having issues on that ballot, had control measures pending in one fashion or another. These include the following: the Delaware legislature will, if it again approves legislation this year, revise the state constitution to forbid the state from spending more than 98% of estimated revenue plus the unencumbered carry-over balance. The Massachusetts legislature passed a tax limitation amendment in 1978 which, if also approved by the 1979 legislature, will be subject to voter approval in 1980. In New York, a tax limitation amendment was introduced in 1975 and has been pending since then. And in Washington, several initiatives must be enacted into law by the 1979 legislature or be referred to the voters in the next general election (Tax Administrators News, 1978). Furthermore, a property tax-limiting referendum in South Dakota presumably will be on the 1980 ballot—a requirement that initiatives be filed a year prior to the general election kept it off the 1978 ballot (Stanfield, 1978: 1821). Since 1979 is a heavy year for state legislatures (annual, biennial, and long sessions), it is not unreasonable to expect even more limiting activity.[9]

There are, however, some approaches that governments—particularly urban—could consider in the effort to maintain the traditional budget process. First, reduced property taxes on housing may be effective in blunting the limitation drive. Evidence from 1978 suggests that high effective property tax rates are associated with rebellion of either the tax or spending limit varieties. Measures removing property from the tax base and thus driving the rate up on parcels remaining in the base may be conducive to rebellion. Some device to constrain by law tax bills during general reassessment may also be an appropriate measure to preserve the equity adjustments produced by revaluation while preventing dramatic local budget increases. Second, existing controls—not just rate limits—may be effective in blunting rebellion. Control initiatives in both Nebraska and Colorado failed, possibly because of existing statutory controls. Such controls have the advantage of being easier to alter than are constitutional amendments and may be subject to more careful drafting than popular initiatives might be. Thus, municipalities may find it attractive to work for statutory controls as a more congenial limiting device than constitutional initiatives. Of particular significance is the extent to which alternative sources may legally be used to maintain services demanded by citizens. In other words, some limits may foreclose spending increases financed by any means, including user charges, nonproperty taxes, or intergovernmental assistance; others permit substantial freedom to increase spending from nontax sources. Surely greater freedom is to be preferred by urban governments.[10]

While the climate is not conducive to speculation about the appropriateness of control initiatives, some consideration may be useful. This voter-driven mechanism can offer taxpayer groups a response avenue when the existing political structure seems to ignore their problems, thus releasing special political pressures in society. Of course, the mechanism does produce some peculiar problems. The proposals are frequently poorly drafted, with the result that they may conflict with existing law. As an example, the Idaho initiative has been ruled in conflict with the state constitution. The proposals also may produce unanticipated effects, as with the California Proposition 13 effect, which, by producing dramatic tax bill increases when a property sells, may severely hinder family mobility. While the strange structural consequences and legal complications of the voter-driven controls may be inconvenient, these perversions can be accommodated. On the less pragmatic level, there are doubts concerning the social desirability of single-issue voting: The required majority of those voting can inflict severe costs on the rest of society with dramatic consequences for the social fabric. Their major disadvantage must emerge from the absence of minority power in the direct legislation system. An initiated referendum has no provision for executive veto, creation of political stalemates in the legislative process, or changed negotiating positions, in committees, all vital portions of lawmaking which can serve to protect minorities. If governments are to be representative, direct voter controls are an uneasy adjunct.

NOTES

1. Ecker-Racz (1970: 117-122) discusses state and local borrowing limitations. The Advisory Commission on Intergovernmental Relations (1977: 83-96) presents a state-by-state tabulation of these constraints.

2. Rate structures may be defined as the complex of statutory or nominal tax rates which combine with the exclusions, deductions, and exemptions to define tax liability. Revenue from a given tax base may obviously be increased by increasing the statutory tax rate or by reducing the generosity of the other provisions. Changes in both statutory rates and other portions of structures require legislative action and are within historical control power.

3. Such impacts are examined by Sunley and Pechman (1976). Thirty-two states plus the District of Columbia employed graduated personal income tax structures in 1978 (Commerce Clearing House, 1978).

4. The Watson Amendment initiatives of 1972 and 1976 in California and Proposal C of 1976 in Michigan.

5. As reported in telephone conversation with the Massachusetts Secretary of State Office.

6. Opinion of the (Idaho) Attorney General No. 78-37, September 15, 1978.

7. Similar patterns emerge for state-local spending combined.

8. Urbanization data are from the 1970 census (U.S. Bureau of the Census, 1973) and the census definition of urbanized population was employed.

9. Kentucky, a state not scheduled to hold a 1979 session, called a special legislative session to consider tax (especially property) controls.

10. McCaffery and Bowman (1978) review the early responses to California Proposition 13, a possible guide to government reactions elsewhere.

REFERENCES

Advisory Commission on Intergovernmental Relations (1977a) Significant Features of Fiscal Federalism, 1976-77 edition, Vol. 2: Revenues and Debt. Washington, DC: author.

--- (1977b) State Limitations on Local Taxes and Expenditures. Washington, DC: author.

Commerce Clearing House (1978) State Tax Handbook as of October 1, 1978. Chicago: author.

ECKER-RACZ, L. L. (1970) The Politics and Economics of State-local Finance. Englewood Cliffs, NJ: Prentice-Hall.

McCAFFERY, J. L. and J. H. BOWMAN (1978) "Participatory democracy and budgeting: the effects of Proposition 13." Public Admin. Rev. 38(November/December): 530-538.

MIKESELL, J. (1978) "The significance of reassessment cycles for property tax performance." Presented at the annual meeting of the Allied Social Science Association, Chicago, August 30.

National Journal (1978) "The November 'tax revolt' background" (November 11): 1783.

SHANNON, J. and C. S. WEISSERT (1978) "After Jarvis: tough questions for fiscal policymakers." Intergovernmental Perspective 4(summer): 8-12.

SMOLKA, R. G. (1978) "Election legislation," pp. 229-259 in P. Albright (ed.) The Book of the States, 1978-79. Lexington, KY: Council of State Governments.

STANFIELD, R. L. (1978) "Less than a full-scale rebellion." National J. (November 11): 1820-1821.

State Tax Review (1978 issues) Commerce Clearing House.

STRUM, A. L. (1978) "State constitutions and constitutional revision, 1976-77," pp. 193-213 in P. Albright (ed.) The Book of the States, 1978-79. Lexington, KY: Council of State Governments.

SUNLEY, E. M., Jr. and J. A. PECHMAN (1976) "Inflation adjustment for the individual income tax," pp. 153-166 in H. J. Aaron (ed.) Inflation and the Income Tax. Washington, DC: Brookings Institution.

Tax Administrators News (1978) "Many tax limitation proposals to appear on November ballots," 42 (October): 109-110, 120.

U.S. Bureau of the Census (1978) Government finances in 1976-77. Washington, DC: U.S. Government Printing Office.

——— (1968) Government finances in 1966-67. Washington, DC: U.S. Government Printing Office.

——— (1973) County and City Data Book 1972. Washington, DC: U.S. Government Printing Office.

7

Public Opinion, the Taxpayers' Revolt, and Local Governments

F. TED HEBERT
RICHARD D. BINGHAM

☐ UNQUESTIONABLY, THE "TAXPAYERS' REVOLT" has been one of the most discussed political issues of 1978, occupying the concerns of politicians, political scientists, and public opinion analysts.[1] The passage of California's Proposition 13[2] on June 6, 1978, brought dire predictions of a nationwide revolution with massive restrictions on the revenue-raising powers of state and local governments (e.g., Shogan, 1978). Through initiative and referenda procedures, as well as legislative actions there were tax reduction or limitation proposals on the November 1978 ballots of a number of states. Many defenders of local government were clearly worried. Vernon Jordan, President of the National Urban League, believes that there is a mood of antisocial negativism (he terms the mood: "The New Negativism") creeping through the structure of American society. As Jordan sees it, the keystone of the New Negativism is the taxpayers' revolt—California's Proposition 13, the various states' tax and spending limitations, the Kemp-Roth bill to cut federal taxes, and the proposals for a constitutional amendment to bar Federal deficits. Jordan warns that The New Negativism, although cloaked in "populist slogans and conservative

AUTHORS' NOTE: *There has been an equal contribution by both authors.*

banners," is actually "a reactionary counter-revolution against positive social change and against the efforts of black people and other minorities to make their way into the American mainstream." The New Negativism, in Jordan's view, thus questions the role of government in seeking social change and a revitalization of our cities (Jordan, 1978).

In our view, the November 1978 election results failed to clarify the implications of the taxpayers' revolt for local governments—the units bearing the brunt of the attack on the property tax. The concerns of some that the financial base of local government would be undercut proved exaggerated at least until state surpluses are depleted. But belief that the revolt would be limited to just a few states and localities also proved groundless. Statewide referenda approving limits on taxing or spending passed in twelve states, but were defeated in four. A summary of these referenda are found in Table 9 of the Mikesell chapter in this volume.

The election left many politicians undecided whether to join the revolt, lead it, or oppose it. Political scientists could not tell from the returns whether or not the revolt marked a major shift in Americans' generalized support for the current regime. And public opinion analysts remained uncertain whether it was indeed of national proportions.

Assessing the revolt's impact on local government requires more than a cursory look at election returns; it requires an examination of its institutional setting and of public attitudes toward government and taxes. Politicians and researchers alike can profit from expanding knowledge of the attitudes underlying the revolt. While evidence is still being assembled, enough is in to draw some very important, if tentative, conclusions. We thus examine public attitudes about taxing and spending, examine some possible causes of the "taxpayers' revolt," and speculate about the meaning of the November 1978 elections.

THE COMING OF THE REVOLT

Because it has become a public issue so recently, there has not yet been extensive research from which to sketch the revolt's development. But an important start can be made by looking at the record of public opinion about taxation. Strange as it seems, this record is not a detailed one. Although polling organizations have asked questions about particular tax changes under consideration (e.g., the surtax of the 1960s), few have carefully traced broader tax attitudes.

There is, however, a survey question that the Louis Harris organization has asked each year since 1973—and in 1969: "From your personal standpoint, please tell me, for each tax that I read off to you, if you feel it is too high, too low, or about right.[3] Examination of responses can focus on the proportions of the population that said "too high." Table 1 shows these percentages for each tax, each year. The federal income tax leads as the one most often designated too high; it topped the list in every year except 1973 when it was surpassed by the local property tax. Least objectionable is the state income tax; except in 1973, each year has seen fewer than 50% designating it too high. Finally, the one trend evident is a decline in the proportion viewing the state sales tax as excessive.

Close examination of data in Table 13 fails to reveal evidence of the sharp change in attitudes toward these taxes that might precede a revolt. To summarize the data, tax objection scores were calculated for each year. These are, simply, the average percentage designating taxes too high in that year. (The averages are calculated across all four taxes.) While these scores have little intrinsic meaning, they do aid comparisons across years. The scores are:

1969	1973	1974	1975	1976	1977	1978
57	60	57	56	59	59	56

Ranging from 56 to 60, and failing to show a trend either up or down, these scores indicate that changes in attitudes toward particular taxes—even all of these taken together—do not seem to form the basis of the

TABLE 13: PERCEIVED LEVEL OF MAJOR TAXES

	Too High (percentages)						
Tax	1969	1973	1974	1975	1976	1977	1978
Federal income tax	66	64	69	72	73	69	70
Local property tax	62	68	56	55	61	66	64
State sales tax	60	56	59	53	55	51	45
State income tax	40	53	44	45	48	49	47

QUESTION: "From you personal standpoint, please tell me, for each of the taxes I read off to you, if you feel it is too high, too low, or about right."
SOURCE: "The Tax Revolt," *Public Opinion* 1 (July/August 1978): 29, from surveys by Louis Harris and Associates.

present revolt. The level of feeling that they are too high may itself be high, but it was just as high nine years ago.

An alternative possibility is that there has been growing objection to government spending. Perhaps the revolt has been triggered by a spreading perception that public spending for specific programs is excessive. Since 1973 the National Opinion Research Center has asked:

> We are faced with many problems in this country, none of which can be solved easily or inexpensively. I'm going to name some of these problems, and for each one I'd like you to tell me whether you think we're spending too much money on it, too little money, or about the right amount.

Table 14 shows, for each problem area, the percentage responding "too much." Foreign aid and welfare clearly led throughout the period. The most significant change was the seventeen point jump for welfare between 1975 and 1976. A clear trend has been the steady decline for military spending.

Again, the data can be summarized by averaging, computing spending objection scores. These are:

TABLE 14: OBJECTION TO SPENDING LEVELS

Program (Problem) Category	Spending Too Much (percentages)					
	1973	1974	1975	1976	1977	1978
Foreign aid	70	76	73	75	66	67
Welfare	51	42	43	60	60	58
The military: armaments and defense	38	31	31	27	23	22
Improving the conditions of blacks	22	21	24	25	25	25
Solving the problems of the big cities	12	11	12	20	19	19
Improving the nation's education system	9	9	11	9	19	11
Improving and protecting the nation's heath	5	5	5	5	7	7

QUESTION: "We are faced with many problems in this country, none of which can be solved easily or inexpensively. I'm going to name some of these problems, and for each one I'd like you to tell me whether you think we're spending too much money on it, too little money, or about the right amount."
SOURCE: "The Tax Revolt," *Public Opinion* 1 (July/August 1978): 32, from surveys by National Opinion Research Center, General Social Survey.

TABLE 15: BREAKING POINT ON TAXES

	1969	1970	1974	1977	1978
			(percentages)		
Reached breaking point	54	60	54	66	66
Not reached breaking point	34	29	40	26	30
Not sure	12	11	6	8	4
Total	100	100	100	100	100

QUESTION: (1969 and 1970): "Do you feel that the amount of taxes you are now paying has reached the breaking point or don't you feel that way?"
QUESTION: (1974, 1977 and 1978): "As far as you and your family) are concerned, do you feel you have reached the breaking point on the amount of taxes you pay, or not?"
SOURCES: Release, The Philadelphia *Inquirer* (April 16, 1970) © Chicago *Tribune*; and Louis Harris, "Tax Revolt Not a New Phenomenon," © 1978 The Chicago *Tribune*.

1973	1974	1975	1976	1977	1978
30	28	28	32	30	30

Changes have not been marked. In total, the greater concern for welfare spending has been offset by decreases in the military and foreign aid categories. As with concern over taxes, looking at attitudes toward spending in separate program categories gives little evidence of the sharp change that could spark a revolt.

Approaching tax-related attitudes more directly, and in a very personal way, Harris has used the question shown in Table 15. It reveals expanding dissatisfaction with tax levels. The two most recent surveys show almost two-thirds of respondents indicating that they have "reached the breaking point." The only change between 1977 and 1978 was a decline in the proportion saying "not sure" and a corresponding increase in those indicating that they have not reached the breaking point. In fact, the "not sure" category has shrunk to its lowest level, probably an indication of the present salience of the tax issue.

A quite different, but complementary, question was used by the Advisory Commission on Intergovernmental Relations (ACIR) in its surveys of 1975, 1976, and 1977. Unfortunately, it was dropped in that of the present year. Shown in Table 16, it yielded no evidence that a revolt was likely. In 1977 over 50% favored keeping taxes and services where they were, and this was an increase over the previous two years.

The seeming discrepancy between these two survey findings may be explained by the public's failing to accept the tax-service linkage

TABLE 16: TAX AND SERVICE LEVELS

	1975	1976	1977
		(percentages)	
Keep taxes and services about where they are	45	51	52
Decrease services and taxes	38	30	31
Increase services and raise taxes	5	5	4
No opinion	12	14	13
Total	100	100	100

QUESTION: "Considering all government services on the one hand and taxes on the toher, which of the following statements comes closest to your view?"
SOURCE: Advisory Commission on Intergovernmental Relations, *Changing Public Attitudes on Governments and Taxes, 1977* (Washington, DC: U.S. Government Printing Office, 1977).

implied by the ACIR question. That is, the question leaves no option for the respondent who wishes to keep services "about where they are" but "decrease taxes"—and who sees this as feasible. Evidence that a sizable proportion of the population might hold this view can be found in the results of several polls taken shortly after the passage of California's Proposition 13. A number of national polling organizations asked respondents whether they would vote for a measure like Proposition 13 if given the opportunity. The three noted below included in their questions references to opponents' contentions that adoption might result in service cuts. Their findings were:

	Would Vote for
NBC and Associated Press (1978)	53%
CBS News-New York Times (1978)	51%
ABC News-Harris Poll (1978)	63%

The ABC-Harris Poll study found strong support despite its question containing a warning that cuts in local government services might be as high as 35%. Significantly, though, there was a follow-up question: "Would you favor or oppose such a 57% cut in local property taxes if it meant cutting the amount your local government spent on (READ LIST) by 35%?" This series of questions yielded sharply different results. With a deep cut in services introduced as a likely prospect, the percentage saying they would vote favorably declined markedly, as Table 17 shows. Reconciling these findings requires a conclusion that the public does not see such cuts as likely. Indeed, there is some direct

TABLE 17: TAX CUT AND PROGRAM SPENDING CUT

	Favor	Oppose (percentages)	Not Sure
Police protection	30	62	8
Fire protection	26	66	8
Collecting garbage and trash	42	48	10
The number of teachers in the public school system	35	57	8
Educating children in the public schools	27	66	7
Maintaining and repairing roads	38	56	6
Service in public hospitals and health care	31	62	7
Aid to the elderly, disabled, and poor	24	71	5

QUESTION: "Would you favor or oppose such a 57% cut in local property taxes if it meant cutting the amount your local government spent on (READ LIST) by 35%?"
SOURCE: ABC News-Harris Poll on Proposition 13, conducted June 15-17, 1978.

evidence of this. Data in Table 18 reveal that a majority of respondents think a 20% cut in local property taxes would not mean seriously reducing services or raising taxes. In fact, a survey conducted by *Newsweek* and the American Institute of Public Opinion (Gallup) found no local services for which a mjority indicated that too much money was being spent. Forty-two percent said this was so for social services (welfare, counseling, mental health, etc.), but only one-fourth said so for public schools and even fewer for recreation, public hospitals, road repair, police, sanitation, library, and fire department. In all cases except social services, the category "right amount" received a plurality or majority ("The Tax Revolt," 1978, p. 33).

WHY LOCAL GOVERNMENT—WHY THE PROPERTY TAX?

The two questions heading this section are appropriate even though activities associated with the revolt have not been limited to local governments and the property tax. In fact, of the referenda considered in the November election, only eight bore directly on them, while the other votes concerned state taxes or state spending. Yet, the importance of California's Proposition 13 and the emergence of local-level revolts (not reflected in data on statewide referenda) make it important to ask just why this particular tax is getting so much attention.

TABLE 18: TAX CUTS POSSIBLE WITH NO SERVICE REDUCTION

Level of Cut	Percentage of Respondents Choosing
Not at all	7
Under 10%	5
10% to 20%	15
20% to 35%	24
35% to 50%	17
50% to 75%	9
75% to 100%	2
Not sure	21
Total	100

QUESTION: "How much do you think it is possible to cut local property taxes before it would mean making serious reductions in local government services or raising other kinds of taxes—100 to 75%, 75 to 50%, 50 to 35%, 35 to 20%, 20 to 10%, under 10%, or not at all?"
SOURCE: "The Tax Revolt," *Public Opinion* 1 (July/August 1978): 33, from survey by Louis Harris and Associates, June 15-17, 1978.

Data in Table 13 revealed that the public chooses the federal income tax as being "too high" to a greater extent than it does the property tax. Further, there is evidence that, among units of government, it is the federal—not the local—that is particularly disliked. A June 1978 NBC and Associated Press poll asked, with regard to each of four types of government, "Do you feel you get your money's worth from the tax dollars you pay . . . or don't you think you get your money's worth?" Table 19 shows that local governments fared better than federal or state. For local *schools*, a plurality of respondents indicated that they do get their money's worth. Other recent investigations confirmed this finding. Peculiarly, one study contradicts it. The Advisory Commission on Intergovernmental Relations (1978) used the following question: "From which level of government do you get the most for your money—federal, state, or local?" It found that 35% chose federal, 26% local, and 20% state. This order has persisted in annual surveys since 1972. In its 1978 report the commission notes that its findings for the present year differ from other surveys and suggests this results from either timing of the surveys (ACIR's was before passage of Proposition 13, the others after) or question wording. The most significant difference in wording is that ACIR's question does not refer to "tax dollars" as do two of the contradictory polls. (The third asked which level

TABLE 19: PERCEIVED MONEY'S WORTH FROM TAX DOLLARS

| | Unit of Government | | | |
| | Federal | State | Local | Local Schools |
			(percentages)	
Get money's worth	21	30	39	45
Don't get money's worth	73	63	53	44
Not sure	6	7	8	11
Total	100	100	100	100

QUESTION: "Do you feel that you get your money's worth from the tax dollars you pay to the Federal Government, or don't you think you get your money's worth?" (Successive, similar questions were asked regarding other units.)
SOURCE: NBC News and Associated Press Poll, June 12-13, 1978.

wastes the biggest part of its budget, and the order was federal, state, local.) It seems reasonable that, in the climate of the revolt, use of the word "tax" may have been critical in producing the response differences. One point on which all four surveys agree, though, is that local governments are viewed more favorably than state governments. Yet, it is the local governments' source of revenue that has been most severely under attack (although displeasure with state spending seems to be spreading).

Characteristics of institutional arrangements provide an explanation for this anomaly of major attention focusing on what may be the most trusted units of government. Petition drives and threats of petition drives have been important tools of revolt leaders. Proposition 13 resulted from a successful initiative effort. California is one of seventeen states that permit popular initiation of constitutional provisions. Six others provide the initiative process for securing referenda on legislative items. (An additional eleven states provide the initiative procedure for some local governments–Council of State Governments, 1978: 210, 243.) Of course, there is no such mechanism available for attacking the federal income tax.

Availability of the initiative process and dislike of the property tax (the second tax in order of being seen as "too high") combined to make local governments the target. Even though they fare relatively well in the public's evaluation, local governments impose a relatively unpopular tax. Further, it is a tax that is perceived by many as rising. A *Newsweek*-American Institute of Public Opinion (Gallup) poll showed that 58% perceive the amount of real estate tax they pay to have increased

TABLE 20: SCHOOL BOND APPROVAL RATE

Year	Elections Held	Percentage Approved
1965	2041	75
1966	1745	72
1967	1625	67
1968	1750	68
1969	1341	57
1970	1216	53
1971	1086	47
1972	1153	47
1973	1273	56
1974	1386	56
1975	929	46
1976	770	51

SOURCE: U.S. Department of Health, Education, and Welfare, National Center for Education Statistics, *Digest of Education Statistics, 1977-78* (Washington, DC: U.S. Government Printing Office, 1978), p. 67.

"a great deal" in the past few years. This compares to 25% for sales tax, 36% for federal income tax, and 15% for state/local income tax. The same survey found that only 10% identified the real estate tax as the fairest form. ("The Tax Revolt," p. 29). Approaching from the opposite direction, the ACIR survey asked respondents to identify the least fair tax, and the property tax led all others (ACIR, 1978). So, there is a combination of widespread perception that the property tax is climbing most rapidly and that it is least fair among major tax sources.

Finally, there is a long tradition of citizen participation in setting local tax policy—and of state constitutional regulation of that process. In fact, tax politics is differentiated from other policy arenas by the significant role the mass public plays. Five states require referenda on all property taxes levied by some types of local governments, and thirty-four more require them before state-imposed tax limits can be exceeded (ACIR, 1977: 152-167). Similarly, of the twenty-nine states authorizing local government use of the general sales tax, sixteen require referenda for its imposition (ACIR, 1977: 188-189). Other taxes and charges are also submitted to voters in some localities, either by choice of the governing body, in accordance with state law, or because of local charter requirements.

Further, in many states decisions by local governments to raise revenue through sale of bonds require voter approval. With regard to

TABLE 21: "TAXES" AS MOST IMPORTANT PROBLEM FACING STATE

State	Taxes—All Mentions	Region	Date of Interviews		
State A	70%	Northeast	August/September	1978	(personal)
State B	44%[a]	West	August	1978	(telephone)
State C	40%	Northeast	July/August	1978	(personal)
State D	38%	Northeast	July/August	1978	(personal)
State E	35%	Midwest	August/September	1978	(telephone)
State F	26%	Northeast	August	1978	(telephone)
State G	19%	Midwest	June/July	1978	(telephone)
State H	19%	Northeast	June	1978	(personal)
State I	15%	Midwest	August	1978	(telephone)

QUESTION: "What do you think are the most important problems facing your state at the present time?"
[a]*QUESTION WORDING*: "What do you think should be the top priority for the next governor of _____ to work on?"
SOURCE: Andrew J. Morrison, Market Opinion Research of Detroit, Round Table on Taxpayers Revolt, Annual Conference of the Midwest Association for Public Opinion Research, Chicago, Il, October 19-21, 1978.

these particular actions there is a bit of data that suggests the long-term development of the taxpayers' revolt. The passage rate of public school bond issues has sharply declined. Despite far fewer issues being submitted to the voters (see Table 20), the rate of passage has dropped. From the fairly stable approval rate above 70% in the 1960s, there was a drop to a low of 46% in 1975. Unfortunately, data are not available to determine whether a similar pattern exists for referenda on other local revenue issues.

A LOCALIZED PHENOMENON?

We are, however, not convinced that there is a nationwide revolt. Rather we see increasing evidence that the taxpayers' revolt should be examined on a state by state basis. Early evidence of this contention was presented at the 1978 Conference of the Midwest Association for Public Opinion Research by Andrew Morrison of Market Opinion Research (Morrison, 1978). Morrison reported the results of preelection polls done for candidates in nine states. Early in the questioning, respondents were asked the open-ended question: "What do you think

TABLE 22: REACTIONS TO PROPOSITION 13 CONCEPT

State	Total	Responses Yes, Is Needed	No, Is Not Needed	Don't Know	Date of Interviews	
State A	100%	73%	16%	11%	August/September	1978
State B	100%	68	17	15	July/August	1978
State C	100%	65	21	14	July/August	1978
State D	100%	62[a]	25	14	August	1978
State G	100%	64	23	13	June/July	1978
State I	100%	48	36	15	August	1978

QUESTION: "You may know, a tax proposition recently was passed by the voters in California which substantially reduced property taxes and put a limit on future increases. Do you think such a plan is or is not needed in your state?"
[a] "As you may know, California voters recently passed Proposition 13, a property tax limitation proposal which would cut property taxes by 60%, though other types of taxes could increase. Would you favor or oppose a similar limitation of property taxes in your state?"
SOURCE: Andrew J. Morrison, Market Opinion Research of Detroit, Round Table on Taxpayers Revolt, Annual Conference of the Midwest Association for Public Opinion Research, Chicago, Il, October 19-21, 1978.

are the most important problems facing your state at the present time?" Respondents were thus afforded the opportunity to name as many "important problems" as they wished. Table 21 shows the percentage of respondents in each of the nine states mentioning "taxes" as one of these important problems. The results are quite startling and do indeed suggest that the revolt is a localized phenomenon. Taxes were mentioned by 70% of the respondents in state A but by only 15% of the respondents in state I. In general, the respondents in the nine states were not overly concerned with taxes. In only three states did 40% or more of the respondents mention taxes.

In six of the same nine states Market Opinion Research asked a specific question regarding the need for a Proposition 13-type tax reduction. The results, shown in Table 22, indicate clear support for such a measure. Even in state I, 48% of the respondents indicated support for a tax reduction measure. Apparently, respondents in state I generally favor a tax reduction although they do not consider taxes to be an important problem. What we do not know, however, is how the voters in state I would actually vote (or did vote) on the issue, since we are unable to identify the state, (although Morrison did identify regions). Our feeling is that, in many cases, voters are much more likely to voice complaints about the tax system that they are to actually vote for a proposition 13-type measure.

Michigan is a case in point. Michigan voters faced two initiative proposals on the November ballot. Proposal J was similar to California's Proposition 13. It would have reduced real and personal property tax assessments from 50% to 25% of the cash value of property; limited state equalization increases to 2.5% for any year; established a maximum rate of 5.6% on the state income tax; authorized a 1% income tax for school districts; and prohibited the legislature from requiring new or expanded local programs unless fully funded by the state. The other proposal, Proposal E, was much more limited. Proposal E also would have prohibited the state from adopting new or expanded local programs without full state funding. In addition, it would have limited state taxes and revenues to its current proportion of total state personal income except under emergency conditions; prohibited a reduction of state aid to local governments; and prohibited local tax increases without voter approval.

Prior to the election it appeared that both proposals would pass. For example, 57% of the respondents to an October 2-10 Michigan poll (Market Opinion Research) answered "yes" to a question asking if Proposal J should be adopted (Bryant, 1978). Proposal J did not pass, however. Instead the voters approved the milder Proposal E. While preelection polls certainly cannot be confused with election results, or in some cases with election predictions, it did appear that the proposal would pass. But why did it fail? We suspect that when push came to shove, Michigan voters were not willing to approve harsh restrictions on government although they were willing to strike out at government through the polls.[4]

Three other states—Idaho, Nevada, and Oregon—had Proposition 13-type measures on their ballots. The measures passed in Idaho and Nevada but failed in Oregon. The Oregon measure would have frozen the full cash value of property at the 1975-1976 level and would have limited the taxes on real property to 1.5% of the full cash value. The Idaho proposal established 1978 as the base year in determining actual market value and restricted property taxes to 1% of actual market value, while the Nevada proposition limited assessments to the purchase value (NACO, 1978). While the 1% limit on property taxes in Idaho is a real rollback from the current property tax rate of 1.7%, it is not indicative of a national trend. Nor is the antitax "victory" in Nevada very meaningful. Nevada is an unusual case. For one thing nevada had a projected operating surplus of 17% of revenues over expenditures. For another, Nevada voters must approve the measure again in 1980 before it can become effective. Nevada also has another unusual source of income—taxes on gambling—not available to most states. Given these facts we are not convinced that the vote means much to the rest of the nation.

There is another reason to question the idea of a taxpayers revolt. In 1978, sixteen states had projected operating surpluses in excess of 6% of expenditures as is shown by Table 23. Many of these states (50%) had measures on the ballot designed to limit spending or cut taxes at the state and/or local level. In Arkansas, Nebraska, and Oregon these measures lost—in spite of substantial surpluses. Limits won in North Dakota, Texas, Nevada, South Dakota, and of course, earlier in California.

TABLE 23: STATE GOVERNMENTS WITH OPERATING SURPLUSES IN FISCAL YEAR 1978 IN EXCESS OF 6 PERCENT OF EXPENDITURES[a]

State	Projected Surplus (in millions of dollars)	Projected Surplus (percentage of 1978 expenditures)
Alaska	570.1	66.5
North Dakota	157.4	57.2
Arkansas	189.3	27.9
Utah	53.9	20.1
Texas	622.6	20.0
California	2,157.0	17.6
Nevada	36.9	16.9
Wisconsin	271.1	13.8
Kansas	117.8	13.8
South Dakota	21.3	12.9
Montana	24.1	11.3
Oregon	107.0	10.5
Nebraska	50.8	10.5
Vermont	17.5	9.6
Indiana	110.8	7.3
New Mexico	40.9	7.0

[a]These figures represent fall 1977 projections of what budget positions would be several months later. As such, they differ from actual outcomes. For example, the surplus shown for California is substantially less than that currently reported by the state.
SOURCE: National Association of Counties (1978) The Tax Reform Primer II (September), p. 25.

Among states without large surpluses, only nine (26%) had measures on the ballot. All but one of these passed (Colorado), but most were much less extreme than Proposition 13: Massachusetts authorized differences in assessment rates for businesses and homeowners; Missouri authorized decreases in tax rates when property is reappraised; Michigan (as shown above) passed the less extreme of two measures; and Illinois passed a purely advisory item. Only Idaho adopted a Proposition 13 look-alike.

The results are thus simply inconclusive. Some of the states with the largest projected surpluses imposed no additional restrictions on revenues nor passed expenditure limits. Even in the cases where restrictions won, it is difficult to conclude that there is a tax revolt. North Dakota, for example, with a 57% surplus, voted to reduce personal income tax rates and increase the corporation tax. Is this a tax revolt or merely an adjustment to deal with a substantial state surplus? We suspect it was the latter.

Table 23 includes Wisconsin as one of the states having a substantial ($271 million) surplus. Wisconsin, has no initiative process, and there was no proposal on the ballot. This, however, did not stop the surplus from becoming the main issue of the 1978 gubernatorial campaign. Apparently, both candidates concluded that there was a sizable group in Wisconsin who actively desired reductions in both taxes and service levels. Thus one candidate focused his campaign on the use of the surplus for property tax reductions while the other emphasized a reduction in the individual income tax. The surplus issue was clearly the dominant campaign issue. A recent study, however, concluded that there was no antitax fever in the state. The Wisconsin Tax Reform Commission completed a random telephone sample of 1,016 Wisconsin residents during the period July 15 through September 10, 1978. Based on their survey results they concluded that no more than 20-25% of the population in the state desire tax and service reductions, and these are strongly counterbalanced by others who disapprove of expenditure reductions or favor increased services (Kinney, 1978). The tax issue in Wisconsin, at least in terms of the gubernatorial campaign, seems to have been overblown.

CONCLUSION

It is impossible to deny that we have seen a trend toward increased restrictions on state and local government spending powers. Since 1970 at least fourteen states have placed restrictions on the power of local officials to raise property taxes (Shannon and Weissert, 1978). No doubt one explanation lies in the impact of inflation on governments (which raised assessments and rates to keep up) and on individuals who saw taxes as, perhaps, the only charge over which they might exercise some control. There is a longer-term development, though, that must not be overlooked. In recent years there has been growing displeasure with governments and their activities—even at the same time as citizens continue to support specific programs.

Studies by the Survey Research Center reveal that an increasing proportion of the population believes that government wastes "a lot" of money. In 1964, only 47% selected this response (Survey Research Center, 1971), but it has expanded steadily until today, according to a CBS News and New York *Times* poll (1978), 78% think so. Simultaneously, there has been a declining trust in government. Whereas in 1958 75% said that they could always or most of the time trust the government in Washington to do the right thing, this had declined to 34% by 1976. While the decline may not have been as sharp for state and local governments, it seems likely that trust in them has dropped, too.

And yet, "the highly touted taxpayers' rebellion didn't happen ... or at least not quite the way proponents of tax rollbacks like California's Proposition 13 had hoped" (McBee, 1978).

The National Taxpayers Union was particularly concerned with the ballot issues in eight states—Arizona, Colorado, Idaho, Michigan, Nebraska, North Dakota, Oregon, and South Dakota. Measures passed in only four—Arizona, Idaho, North Dakota, and South Dakota. Five of these eight states had substantial budgetary surpluses.

Admittedly, we have taken a controversial view of the tax revolt—we do not believe that the revolt is a nationwide phenomenon. Tax issues must be examined on a state-by-state basis.[5] By doing just this we hope that we have presented some reasonable arguments. In 1976 the Wall Street *Journal* declared the "Frostbelt-Sunbelt" issue or the "Second War Between the States" to be "the most overblown political issue of 1976" ("Another round to the sunbelt" (1977): 1037). We suggest that the taxpayers' revolt may be the most overblown political issue of 1978.

NOTES

1. For example, a discussion among representatives of these three groups took place at a roundtable session of the 1978 meeting of the Midwest Association for Public Opinion Research. No effort will be made here to precisely define "taxpayers' revolt"; the expression will be used as it is in the mass communications media.

2. Proposition 13 sets the maximum tax rate on real property at 1% of full cash value; rolls back the assessed property valuation to the 1975-1976 level except when a property is sold or there is new construction; limits future property tax assessments (on which the rate would apply) to an annual increase of not more than 2%, except in the years when a property is sold or there is new construction; allows real property to be assessed at current market value, rather than at the 1975-1976 level, when the property is sold or there is new construction; prohibits the California legislature from levying a state property tax or property transfer tax; prohibits any increase in the property tax even by public vote; prohibits the State legislature from adopting any new or increased taxes except by a two-thirds vote of the members of each house; and requires a two-thirds approval of "qualified electors" for any special taxes levied by local government or special districts.

3. The surveys have included the federal corporate income tax and capital gains tax; however, the percentage responding "not sure" ranged up to 60%. These are omitted here.

4. For this reason we have chosen to ignore the "advisory" votes in Massachusetts and Illinois. While Mikesell in his article in this volume emphasizes the Massachusetts and Illinois votes we view the results as an expression of resentment but are not willing to equate advisory votes with constitutional or statutory measures.

5. Although it predates the present revolt, evidence from Urban Observatory surveys of citizens in ten major cities confirms the cross-state variability of tax attitudes. The percentage thinking taxes were "too high" ranged from 68% (Baltimore) to 27% (Albuquerque and Nashville). Significantly, this ranking of cities correlated well (.94) with their ranking on effective property tax on single family homes (see Fowler, 1974).

REFERENCES

ABC News–Harris Poll (1978) "Release: poll on Proposition 13" (June 19).
Advisory Commission on Intergovernmental Relations (1978) Changing Public Attitudes on Governments and Taxes, 1978. Washington, DC: U.S. Government Printing Office.
––– (1977) Significant Features of Fiscal Federalism, 1976-1977. Washington, DC: U.S. Government Printing Office.
Another round to the Sunbelt" (1977) National J. (July 2): 1033-1037.

BRYANT, B. (1978) "Round table on taxpayers revolt." Annual Conference of the Midwest Association for Public Opinion Research, Chicago, October 19-21.

CBS News and New York Times (1978) "Release" (June 27).

Council of State Governments (1978) Book of the States, 1978-1979. Lexington, KY: Council of State Governments.

FOWLER, F. J., Jr. (1974) Citizen Attitudes Toward Local Government, Services, and Taxes. Cambridge, MA: Ballinger.

JORDAN, V. E., Jr. (1978) "The new negativism." Newsweek (August 14): 13.

KINNEY, K. S. (1978) "No anti-tax fever in state, study finds." Milwaukee J. (November 19): 1, 14.

McBEE, S. (1978) "Prop 13 aftermath: a tax revolt that didn't happen." Washington Post (November 9): A9.

MORRISON, A. J. (1978) "Round table on taxpayers revolt." Annual Conference of the Midwest Association for Public Opinion Research, Chicago, October 19-21).

National Association of Counties (1978) The Tax Reform Primer II.

NBC News (1978) "NBC reports: 'mad as hell'; taxpayers revolt" (June 20).

SHANNON, J. and C. S. WEISSERT (1978) "After Jarvis: tough questions for fiscal policymakers." Intergovernmental Perspective 4 (summer): 8-12.

SHOGAN, R. (1978) "Taxpayers rebel, take the initiative." Los Angeles Times (May 7).

Survey Research Center, University of Michigan (1971) "The 1964 SRC election study," revised edition. Ann Arbor, MI: Inter-University Consortium for Political Science.

"The tax revolt" (1978) Public Opinion (July/August).

8

A Mayor's Perspective on the Tax Revolt

HENRY MAIER

☐ IN JUNE 1965 A GROUP of mayors and urban scientists met to discuss problems of data useful for policy analysis and the possibility of forming urban observatories. Dr. Robert Wood, then of MIT, later the Deputy Secretary of HUD, and now President of the University of Massachusetts, had, in an essay, espoused the idea that we ought to have urban observatories to monitor urban indicators much as we have had agricultural observatories, and that we ought to have a unity of town and gown to achieve this end. Fortunately, after Robert Wood became the Deputy Secretary of HUD, he put the national Urban Observatory in the HUD budget, and we had a very handsome start toward fulfilling the need for intercity comparative research. Unfortunately, its federal funding was cut off during the Nixon administration, data series became problematic, and policy research was retarded.

As the mayor of a large metropolitan city, I have always had a strong interest in the development of hard data regarding the urban world. Good policy decisions cannot be made in an information vacuum. Unfortunately, although more data are available now than fifteen years ago, a particular bit of information that one needs at a given moment is often elusive. Further, the data that are available are often misinterpreted by the media reporting on an issue.

I have found that Finagle's Laws of Information define the situation very well, and may well take their place beside such other well-known rules as Murphy's Law and Parkinson's law. Finagle's Laws of Information are:

(1) The information you have is not what you want.
(2) The information you want is not what you need.
(3) The information you need is not what you can obtain.
(4) The information you can obtain costs more than you want to pay.

Now, to some degree all these laws pertain when we try to predict the effects of retrenchment on local governments. However, I believe that a number of comments are in order based on the information that we already have.

PROPOSITION 13

Proposition 13 has received more publicity than any proposition since Caesar and Cleopatra. There is no denying its seductive appeal to hard-pressed taxpayers. It sweetly whispers, let me take you away from all this; come with me kid, and you'll live on Easy Street where you'll never have to pay for any potholes. Under the spell of Proposition 13, the media raised up a landlord's advocate as a very unlikely folk hero. The phrase, "taxpayers' revolt" has practically become a standing newshead. Day after day, we have new examples spotlighted to prove that the revolt goes on: Tax cut legislation appearing in a number of states; the candidates vying, in my opinion disgracefully, to see who can promise the biggest budget cut of all; a radio personality in New Bedford, Massachusetts, leading a march on City Hall to lower reassessment; and national TV giving coverage to a homeowner in Florida who is bulldozing his home down rather than paying increased property taxes.

Where the cry of the Colonists was "taxation without representation is tyranny," Proposition 13 boisterously contends that taxation *with* representation is also tyranny. The fact is no one has ever loved the property tax since the time it appeared on the scene as one of the first forms of taxation in America. But only recently has it become a *national* issue.

For some central-city mayors it is ironic that they must bear the brunt of the revolt. For many years these mayors have contended that the local property tax is an important factor in the crisis of the cities. The property tax is and long has been overloaded with the costs of too many functions which it was never intended to finance. The property tax was never built to take on the costs of health, education, or poverty. It cannot cope with the social and economic problems of today's central cities. As this tax has climbed higher and higher in the face of these problems, and with the added impetus of inflation, it has contributed to urban unrest. It does not take a Howard Jarvis to let local politicians know that it is the most unpopular tax in America. In my opinion, it is also the most unfair. But for years nobody would listen to this thesis or at least nobody would act to rectify the emerging problem of overreliance upon a regressive tax.

For some reason the property tax did not have the glamour of other issues of the sixties. It was not a top priority of research in urban affairs. It was difficult to impress presidential candidates in the Wisconsin primaries with the fact that it should be a prime concern. At the Democratic Convention in 1972, when I brought up the concern in my address, the TV cameras were all at a confrontation in the lobby of George McGovern's hotel.

Tax Revolt or Counter Revolt? Proposition 13, of course, was not the first rebellion to sweep across the nation from California. I think the rebellion of thirteen years ago out of California had pretty much the same characteristics of media treatment as does Proposition 13. Most people will remember how the flames of Watts spread to city after city—Newark, Detroit, Chicago, Milwaukee. At that time I contended that there was a relationship between the lack of resources of the central city, the overload on the property tax, and the social unrest of our times. Central-city mayors just could not deal with the underlying social problems in the central cities with this lack of resources. Consequently, they were forced to rely upon the property tax. Unfortunately, reducing taxes will not help solve the basic problems that require sufficient financing of urban society. Indeed, the problems are going to become worse.

In a sense, Proposition 13 represents not a revolt but a counterrevolution of the propertied class. It does nothing for the nonpropertied renters, and very little for low-income workers. In fact, these low-income renters are likely to be the first to suffer from cutbacks in services. I believe that this can lead to even greater unrest in our cities in the future.

CONSEQUENCES OF THE SPIRIT OF RETRENCHMENT

The spirit of Proposition 13 can also lead to a cutting back of state and federal programs which are not financed by the property tax and are needed to meet the problems of poverty in our central cities. In recent years it has become most fashionable to criticize the programs of the Great Society, claiming they have failed to meet their goals. Senator Pat Moynihan, however, has pointed out that most of these programs were oversold and, with the coming of the Vietnam War, underfinanced so that they were doomed to failure in meeting their promises. And, in fact, it is likely that we will continue to see an everincreasing Pentagon budget. Congress is already projecting the Pentagon budget four years ahead with increases beyond the rate of inflation. While this is being contemplated, we will see increasing cuts in programs that will help the cities, as was the case in the last Congress when the countercyclical aid to the cities was cut from the budget.

It has become fashionable to say "you can't solve problems by throwing money at them," although a long time ago we mayors decided that money beats the hell out of anything that's in second place. This has certainly not been the Pentagon approach. As the author Irving Howe (1978) reminded us recently in the *New Republic,* "It is also true and right now far more important, that without spending money, social problems cannot be solved."

I might note that you do not have to *throw* money. You can target it carefully in accordance with a well-conceived national urban policy. Howe makes the added point, "We don't even have an adequate idea of what a decent social expenditure would be in America, so impoverished is our collective imagination by the skewed priorities of Pentagon excess, advertising and regressive taxes" (1978: 14).

Not only will cutbacks create additional urban problems, but another of the cold realities likely to face citizens the morning after Proposition 13 is that we really have not saved anything. Sooner or later, the deferred maintenance of urban infrastructure must be faced at an increased cost unless we throw away our older central cities. Where once the old house could have been fixed up with a few shingles and a length of pipe, we may very soon need a whole new roof and a plumbing system. The same concept applies to public projects—roads, sewers, buildings.

Even before the passage of Proposition 13, George E. Peterson (1978) of the Urban Institute, was warning us, "The Capital stock in our cities is wearing out, creating severe maintenance and repair needs ... beyond some point, deferred maintenance and repair necessitate total replacement of capital as the only workable investment strategy." As Peterson pointed out, "this will create new liabilities for a future generation of city taxpayers, as today's deferring of maintenance and repairs becomes tomorrow's need for massive replacement investment." In addition, Peterson noted that we are not investing as much of our budgets in capital expenditures as we were a decade ago: "The capital share of total spending has fallen steadily from thirty percent in the mid-sixties to less than fifteen percent in 1977."

In October 1977, *Nation's Cities,* the magazine of the National League of Cities, also called attention to "the wearing out of urban America" (Nye, 1977: 8). Among other things, that issue noted that America's bridges are falling down. It referred to the sixth annual report of the Department of Transportation which said that there are 105,500 federally aided and city or county bridges that are now deficient. Merely replacing the city and county bridges of America would cost $10.6 billion. In the city of Milwaukee, the estimated costs of replacing deteriorating bridges is $60 million or more. *But Milwaukee is, in contrast with most large cities, on schedule with its entire capital maintenance and replacement.* Streets, sewers, and sidewalks are also wearing out nationally according to the magazine's survey.

More recently, Pierce (1978) suggested that citizens who vote for tax cutbacks may "be living in a fool's paradise." He claimed that

> the reason can be summed up in two drab words; deferred maintenance. Few businesses or homeowners would long tolerate rotting pipes, decaying walls or holes in the ceiling. But such postponements of vital maintenance—delays in repair of aging water mains, sewer lines, storm drains, bridges and roads—are occurring with increasing frequency in cities and states across the country.... Not only do the delays threaten the lifelines of our society, they could also mean much bigger bills in the years to come because the cost of decay is invariably greater than providing timely replacement or maintenance!

Pierce also detailed some of the items that are already coming due in "the heavy bill for years of neglect!" (1) Leaks in Boston's ancient water mains, some dating back to the 1840s, cause the loss of seventy-five million gallons of water in that system daily; (2) New York: two massive water tunnels are crumbling, but neither can be examined or repaired; (3) Houston loses up to 30% of its purified water daily through inadequate pipes. Thousands of water damage complaints pour into City Hall annually as the leaks undermine sidewalks, collapse driveways, and destroy laws. Most of New Orleans's sewers need replacement. Some were purchased, secondhand, from Philadelphia in 1896; in San Francisco, pieces have fallen off of the vaulted Golden Gate Bridge. And to the list we can add the need to remedy the pollution stemming from Milwaukee's antiquated combined sewer system at a cost of more than one billion dollars, while the capital costs of meeting the Illinois federal court's ruling in the antipollution suit may cost, projected with inflation, another one-half billion dollars.

In the final analysis, there are limits on the local response to retrenchment if the municipality is to continue its function as a city. For many years we have been living within the confines of retrenchment. I have long taken the position that there can be no new city programs unless there is financing by other than the property tax. In today's society it is not enough for the local politicians alone to assume the responsibility of retrenchment. Other institutions, particularly the media, must also assume responsibility with accurate and responsible reporting and editorializing. It is easy for those not charged fiscal responsibilities for the city to simultaneous advocate new programs *and* tax cuts. Such activities influence policy and create false expectations within the community. Thus the media have a responsibility to be consistent regarding their overall mix of editorial positions as well as being accurate and consistent in individual articles.

Local experiences have shown that the media do not always assume that responsibility consistently. Before Proposition 13, and against opposition from all the media, I vetoed a paramedic proposal because I believed it to be a Cadillac of the health services. For a number of years I had been resisting it, and I had been hoping that there would be enough money for a preventive health program in the city because we have had some marked success in this area. My veto was overridden, but the points I made in the message are still valid. We simply cannot take on new programs financed by the property tax.

One of the strongest proponents of paramedics was the local newspaper. At least since the beginning of the long tenure of Mayor Hoan fifty-six years ago, the newspaper has focused a critical eye on the obligation of the central city.

Those who criticize the city's fiscal programs often ignore the metropolitan nature of urban government. They may call for cutbacks, but their underlying reasoning is untenable. Cities require more revenues than suburbs. Among the propositions that underlie Proposition 13 are: (1) Only the central city should keep a giant infrastructure intact to service the metropolitan area; (2) only the central city should zone in the poor. As census figures report, the Milwaukee metropolitan area has the most economically segregated suburbs in the United States; (3) only the central city should take the burden and the industrial loss arising out of freeway construction; and (4) only the central city should integrate its school system.

Now these, I believe, have been some of the underlying propositions, from which Proposition 13 has emerged. Tax limitations will have little effect upon suburbs which will normally be able to raise sufficient revenues within the confines of statutory tax rate limitations. The theme then is that only the central city should economize. The conclusion is that the city should cut back on garbage workers. This implies there are no other units of government—only the city. Only 29% of the tax dollar, small in relationship to the total city taxes, is spent by the central city. Metropolitan spending, miniscule in terms of state and federal revenues is even higher yet, but the only target around, apparently, is the central city.

WHO SHOULD PAY?

Nevertheless, the property tax remains a heavy burden mainly because it has been required to finance too many functions for which it was never intended. As I have contended for years, these functions should be removed from the backs of the property tax.

Consider this: Why should the property tax finance education when most Americans spend about 80% of their lives after finishing school in some community other than the one in which they were educated? It seems to me that it would be logical for the state and the federal income tax to pick up these costs.

Or, should not costs brought about by the use of the automobile be paid for by automobile users? Or, should not the social overhead costs of the concentration of the poor in the central city be shared by the rest of the metropolitan area that zones out or other wise excludes the poor?

The number one proposition that faces us is a reform of the tax system to remove the functional burden from the local property tax.

A SPIRIT OF REVIVAL

We also have to try to overcome the kind of cynicism about our cities that is represented by the spirit of Proposition 13. A cynic, as we all know, is one who knows the cost of everything and the value of nothing. He sees the cost but never the benefits beyond his doorstep. Some have termed it a "me first" movement in which everything that is private is extolled over everything that is public. But is it not time that we overcome the "me first" attitude that anything public is inferior? Are public schools necessarily inferior to private schools? suburban schools better than city schools because they are more like private schools? Is private transportation—no matter how wasteful—necessarily superior to public transportation? Is it all right for private buildings to be lavish, while public buildings should be built as cheaply and perhaps as plain as possible, no matter how much we scrimp on their appearance?

Instead of "me first" should we not really be thinking of "us"—all of those who must live together and must work together, and by rights should take a common pride of ownership in their urban community?

And is it not time that we reemphasize the public benefits at the same time that we try to control the private costs of today's city?

REFERENCES

BRYANT, J. Q. (1978) "Laws for the human condition." Guardian (October 22).
HOWE, I. (1978) "The right menace." New Republic 179, 11: 14.
NYE, P. (1977) "The wearing out of urban America." Nation's Cities (October): 8-12.
PETERSON, G. E. (1978) "Capital investment and capital obsolescence." Presented at an urban Institute Conference titled Federal Impacts of the Economic Outlook for Cities, April 5-6.
PIERCE, N. R. (1978) "Eizenstat on urban policy: the problems of politics." National J.

Part III

**Urban Policy Areas:
Prospects for the 1980s**

9

Employment Policy in Postaffluent America

GARY GAPPERT

☐ FUTURE HISTORIANS WILL LIKELY CHARACTERIZE the twenty-five years from 1945 to 1970 in American society as a period of foolish affluence fueled by borrowed money.

During these decades, American society, after years of deprivation during the depression of the 1930s and World War II, attempted to buy back the lost time by becoming the most lavish consumer society in history. The spending spree was financed in large measure by an explosion of personal credit, by federal subsidies for homeownership and highway construction, and through the floating of local bond issues to build the schools, sewers, hospitals, and other infrastructure for the thousands of growing suburbs. At the same time, the decolonization of Europe's overseas territories gave American businesses access to the enormous reservoir of raw materials held by the underdeveloped countries.

The bill for the tremendous splurge is now coming due, and the result is a postaffluent society. The last quarter-century, from 1975 to 2000, will be a period in which American society will be forced to learn to live with new scarcities and to acquire habits of social thrift.

Postaffluent America is a prospect which is one of the legacies of the stormy 1960s. More importantly, the postaffluent prospect is a realiza-

159

tion that the American dream has new limits. This realization began to appear on August 15, 1971 when President Nixon announced his New Economic Program including a wage and price freeze. The events of December 1973 provided further awakening. A federal energy office was created. Voluntary gas rationing was requested and the Persian Gulf oil producers doubled the price of petroleum from $5.11 to $11.65 a barrel. In one sense, awakening was completed in August 1974 when the first president in history was forced to resign from office because the old rules no longer applied. Future shock became present trauma. The Arab oil decision is a symbol of the American loss of control of its economic future. The Nixon resignation is a symbol of a type of personal dislocation which may lurk around the corner of tomorrow for many of us, especially those who serve in the public sector. The passage of Proposition 13 in California in June 1978 is a further manifestation of the public malaise with the quality of governance and management in these United States.

The political actors and the professional managers of the political economy of postaffluent America are now experiencing the realities of retrenchment. There are several different aspects to retrenchment which should be noted.

First, the macromanagement of the national economy no longer has an adequate analytical base to guide it. Keynes's theory is no longer particularly relevant to the prediction of the economic reaction to the traditional tools of policy action: budget deficits and interest rates. The fact is that economists no longer know how to describe and prescribe adequate policies for the national economy.

Second, the power shift in the international political economy means that the United States will increasingly need to export more of its goods and services abroad in order to pay for its use of petroleum and other mineral resources. This means that a growing proportion of our national production will not be available for domestic consumption or redistribution.

Third, the inflationary price increases in such essentials as food, health services, housing and transportation will continue to sharply reduce discretionary income. This reduction contributes to the demands to shrink the size of the public sector.

Fourth, the growth in employment in the late 1970s has been substantial but still inadequate to absorb all the new entrants into the labor force. In 1978, 3.3 million jobs were created with 2.1 million being filled by women, motivated primarily by the economic necessity of providing a second family income.

These and other economic realities of retrenchment will necessitate a major reorientation to the development of national urban and economic policy. In this essay we review the postaffluent prospect. It is suggested that the particular problem of urban youth unemployment has obscured some of the more fundamental problems in the development of a postindustrial labor force. In the 1980s, however, it is likely that education and reeducation policy will become a more integral component of urban economic planning.

THE POSTAFFLUENT PROSPECT

Most futurists are agreed that we are post something, but they disagree as to "what" post something we are. Michael Marien (1973) has tracked over sixty definitions or descriptions for the projected society of the future. The range is from postcapitalist to postwarfare and includes Marcuse's (1974) one-dimensional society and Don Schon's (1973) *Contribution*. These images and metaphors are attempts to bring some kind of thematic unity to the myriad bits of the future emerging in contemporary society. Mumford (1938) said that we are "post-historical," by which he meant that not only are we deficient in a sense of history but that our ignorance is partially excusable because we are post-past patterns.

In a similar context, Robert Jay Lifton (1975) has suggested that we are all survivors and are all post something. As he writes:

> Consider for instance, the widespread inclination to name and interpret the contemporary man or woman not in terms of what he or she might actually be but rather in terms of what has been—that is, in terms of what we have survived. We speak of ourselves as post-modern, post-industrial, post-historic, post-identify, post-economic, post-materialistic, post-technocratic, and so forth. There are pitfalls in this way of managing the present (or the future) after what no longer is (or will be) but the terms have an authentic source in the sense of survivorship, present or anticipated, that so pervades our deepest image of ourselves. [p. 35]

Lifton suggests that as the sense of change intensifies, the boundaries between history and evolution are obscured. People seem to be plunging ahead in an unknowable process devoid of clear destination. They then "suddenly discover, swirling about, the total array of images created over the full course of their historical and evolutionary past." Lifton concludes:

> These images become an elusive form of physic nutriment, to be ingested, metabolized, excreted, and, above all, built upon and recombined in some kind of vital process. [1975: 137]

Some kind of symbolic reconstruction and theoretical integration seems to begin to happen as we treat the future as both a continuous and discontinuous projection or extension of the present and the past. Out of these projections we can fashion a series of mental constructs around which the future can be analyzed. For the purpose of this exposition three constructs have been chosen. These include the notions of postaffluent, postindustrial, and postmachismo.

Each is a meaningful way to organize an analysis of the future which is a projection which builds upon the crisis and circumstances of the late 1960s and the early 1970s. The assumption is that events of the years 1972 to 1974 represent a significant turning point and developmental shift in our society. A sense of crisis peaked in those years. That sense of crisis, although open to evolving interpretations, has not been dispelled by the administration in Washington. Indeed, a particular sense of the crisis is being used by the Carter administration to focus the development of national economic and energy policy.

The crisis of scarcity and the rediscovery of limits has led to an explosion of mass media interpretations. William Ophals (1974) has described it this way:

> Historians may see 1973 as a year dividing one age from another. The nature of the changes in store for us is symbolized by the Shah of Iran's announcement—that the price of his country's oil would thenceforth be $11.87 per barrel, a rise of 100 percent— the challenge is the inevitable coming of scarcity to societies predicated on abundance. Its consequences . . . will be the end of political democracy and a drastic restriction of personal liberty. [p. 47]

This illuminates the issue of postaffluence. The unusual sense of abundance and affluence which flourished from the end of World War

II until the early 1970s is hard to dispel. It has strongly influenced several generations. The baby boom of the same two and a half decades has also created a "bulge generation" which grew and was socialized around the material realities and social illusions of that affluence. There will never again be, in this country, a generation of youth which had it so good. Affluence is now behind us, but we have survived it. It is unclear, however, how its memory will affect us or that unique generation which was nourished into adulthood from a basis of unprecedented material prosperity. The social values of that generation have been a source of continuing discussion in the mass media. When that generation was concentrated on the college campuses or on Vietnam battlefields in the 1960s, it was an easy generation to observe. As this generation enters the workplaces and communities of the adult world, its behavior and values will be less directly observable but will be indirectly manifested in the dynamics of the marketplace and in the evolution of the social economy.

That market place and the myriad transactions of the social economy bring us to the postindustrial characterization. Daniel Bell (1973) has identified the five principal elements of the postindustrial society as he has studied it. These are:

- the change from a goods-producing to a service economy.
- the preeminence of professional and technical jobs.
- the "centrality" of theoretical knowledge as the primary source of innovation and of policy formulation.
- the control of planned technological growth or development.
- the creation of new kinds of computer-based systems for management purposes.

The generation growing into the late twenties in this decade (1970-1980) are entering a postindustrial marketplace where so-called service jobs are the primary economic influence. This service economy demands new kinds of interpersonal relationships among workers, between workers and management, and between workers and consumers. For most people the work experience will have a strong social element. Fewer workers will be oriented toward a machine, or toward the land as farmers or miners. Here again the society has survived the growth of the factories of the industrial expansion which was the dominant economic force in our society from the late nineteenth century through the middle of the twentieth. The service job orientation of the emerging postindustrial period needs to be explored further as we explore the social economy of the future.

But it is also time to note that by 1985 it is projected that almost half of the working force will be women. This is one way to introduce the third postsomething construct. This is the idea that the emerging social system of the next five decades can be usefully characterized as postmasculine, or postmachismo. It is suggested that some of the unique elements of this postmachismo society can be described as follows:

- the strange custom of educating boys and girls together which has now led to joint physical education programs and teams.
- the supervision and direction of young males almost entirely in the hands of females, right through to maturity.
- deemphasis on physical strength especially in the workplace.
- new legal supports to enable those women in the workforce to compete more successfully against men.
- new patterns of sexual permissiveness and romatic love, especially in the emerging adult generation, which reduce the aggressiveness prerogative of men.
- new patterns of lifestyle assertiveness and independence for women.
- a decline in the effectiveness of the military mentality in organizational leadership and structure.
- different criteria of fairness and equity.
- a release of additional creativity as the "new male" emerges and achieves greater acceptability.

The society, in fact, is going to become much more "androgynous" and will have the characteristics of both sexes. What may happen is that a more balanced set of values will predominate in all our institutional arrangements, even in the military. More desirable perhaps, aspects of machismo silliness will no longer have as great an influence in the politics of our society. The most significant example of such masculine silliness of course was the attitudes of both Lyndon Johnson and Richard Nixon toward ending the Vietnam War. A more recent example is the reaction to the Panama Canal treaty.

The cities are on the cutting edge of the emerging social economy whether characterized as postaffluent, postindustrial, or postmachismo.

TABLE 24: AGENDA FOR ECONOMIC PROBLEMS: 1978-1984

Probable Solutions

1. Successful programs for structural unemployment
2. Post-Keynesian policy of sector incentives and selective credit controls
3. Inadequate energy policy
4. A temporary national urban policy
5. Stabilization of welfare programs

Possible Solutions

1. Stabilization of emerging New International Economic Order
2. Reversal of decline in governmental effectiveness
3. Expansion of lifelong learning systems and some institutional integration
4. Successful initiation of national long-range economic planning

Less Likely Solutions

1. Increase in global interventions by multinational and transnational institutions in areas of public safety. Environment, economic planning, etc.
2. R & D support for innovative tinkering with family support systems and nurturing networks.
3. Some kind of decrease population policy with birth licenses, etc.
4. Policy and incentives for social economic consulting including but not limited to job training
5. Utilization breakthrough in solar energy
6. Developmental breakthrough in fusion-based energy

It is the cities, especially the old Northern cities, that have experienced the slowest income growth, the most rapid shift away from manufacturing and toward services, and the largest increases in labor force participation rates among women. Therefore, we can expect policies of economic and employment reform to be focused upon cities.

THE POSTAFFLUENT POLITY

In order to fully understand the possible context for policy development in the next twenty-five years, it is necessary to consider that the evolving national economic crisis is going to force a major and more substantial shift in intergovernmental relations. Just as the New Deal of the 1930s was followed by three decades of refinements and imitations, so it can be projected that the New Federalism of the early 1970s will be followed by three decades of similar refinements and extensions. The New Partnerships theme of President Carter's urban policy builds on the "block grant" and "revenue-sharing" innovations of the previous administration and includes an expanded concern with neighborhoods and voluntary associations. The ultimate New Federalism will be determined by the reality that the national government will increasingly be preoccupied with the management and mismanagement of the national economic system and, more importantly, with the problems of the emerging international economic order.

The agenda of urban economic problems (see Table 24) for the next several years is substantial and although some solutions are probable, other solutions are less likely. These problems will continue to dominate the concerns in Washington, leaving many social and educational problems to the lower tiers of government. The classic use of urban communities as "laboratories" in which thousands of economic and work-related innovations are tried and evaluated offers a good deal of flexibility in the spatial development of economic policy.

At the same time, as the postaffluent constraints become more apparent, it is likely that a higher order of social planning and economic controls will be placed upon the development of both technology and natural resources. As it is, however, the effectiveness of existing public regulations and controls has been so poor that it is increasingly apparent to some observers that the theory and practice of public administration must be reinvested in order to more accurately reflect the postindustrial and postaffluent realities. The public economy resembles a leaky bucket. Resources are drawn from the public well, and are applied to a problem, but most leak away before reaching that problem. New patterns of intergovernmental and interinstitutional relationships will be required in order to recreate the kind of effective public sector needed to manage a steady society in the twenty-first century. In the meantime the social and political environments will be increasingly turbulent.

THE ECONOMIC CONDITION OF POSTAFFLUENCE

There is differing evidence with respect to the economic condition of postaffluence. The assumption has been that affluence means discretionary income and is related to "opportunity, choice, and diversity." The condition of postaffluence would mean, then, that there has been a reduction of limits placed on opportunity, choice, and diversity in the social economy of the United States. In the terms of the economist, postaffluence would be portrayed by a decline or leveling off of real disposable income. When more households have trouble making ends meet and realizing their expectations, they lose "discretion" in a more fundamental sense. In 1973-1975 the expectations of the American middle class were shattered. Within those two years, the purchasing power of the dollar shrank by nearly 20%, the value of household savings and other wealth and assets declined about 12%, and the size of household debts rose by over 18%. According to *Business Week* (March 10, 1975), the total value of all American households' net financial assets at the end of 1974 was only 64% of the value when 1973 began. Furthermore, the stock market decline in those two years wiped out $300 billion in stock values, and the households' equity investments were reduced from $974 billion in 1972 to $495 billion at the end of 1974. Although the misery of the middle class cannot be compared with the subsistence struggles of the poor, the sense of outrage caused by the rather abrupt onslaught of postaffluence in 1973-1974 has been followed by continuing malaise, confusion and puzzlement among all classes of workers.

A decade of stagnation in the incomes of manufacturing workers is something new in the postwar era. In the earlier years, this group made steady gains in earning power, even after erosion by inflation and rising federal taxes. In 1946-1956, for example, a decade that included inflation, recession, and warfare comparable to those factors in a more recent period, the average real take-home pay of manufacturing workers with three dependents rose about 2% per year. In 1956-1966, gains slowed slightly to 1.4% a year. But in the past decade, which by now includes two full years of recovery from the last recession, the gain was a microscopic 0.3% per year.

Even this calculation probably overstates the gain in the average factory worker's purchasing power. Translated into 1976 price levels, the annual gain in the last decade works out to about $5.66 per week. But nearly $4 of this gain goes into higher state and local taxes. For workers in many areas, the rise in local taxation may very well have meant no rise at all in disposable income over the past ten years.

Another characteristic of the beginning of the postaffluent transition is a dramatic increase in consumer debt. Consumer debt rose from 1970 to 1974 by 42% while after-tax personal income in the same period rose by only 37%. This means that a larger portion of postaffluent incomes will be used for debt service and payments on past purchases. Delinquency rates for mortgage loans and installment payments have crept up to about 4% a year. It should also be noted that while consumer debt rose by over 40% in the early years of the 1970s, government spending in the same period rose by less than 30%. Another indicator of the burden of installment debt, measured by the ratio of outstanding installment credit to disposable personal income, is that this ratio rose to over 15 percent in the early 1970s. (In 1945 it has been 1.4%.)

Increasingly, American families have had to rely upon debt to maintain any kind of discretionary spending. In 1976, according to the Bureau of the Census, about 83% of American families had incomes below $25,000. Approximately 11% made less than $5,000. About 20% earned $5,000-10,000, and another 20% earned $10,000-15,000, and that 40% represented the families of the lower middle class—the sub-affluent. Another 32% of American families earned $15,000-20,000. Because of the problems of expectations, it is unclear where in the range of say, $12,500 to $25,000, discretionary spending options begin to emerge and increase. But the dramatic increases in the costs of housing, transportation, medical care, and personal taxes in the first half of the 1970s sharply reduced the discretionary options of the so-called middle class. More than ever before the line between necessities and options depends upon uncontrollable costs and controllable expectations and lifestyles. Ironically, in January 1978 the editors of *Business Week* and other conservative business leaders including former President Ford bemoaned the plight of the 15% of American families earning $25,000-50,000 (another 2% earned more than $50,000). *Business Week,* in analyzing the combined impact of the new Social Security Act and the proposed income tax reduction, indicated that this group which "provides the savings" for American capital investment "is threatened with what amounts to liquidation."

> Take one example, a family of four with $35,000 a year in income. The Social Security bill signed by the President last month will increase this family's payroll tax to $1,070 this year, $2,291 in 1983, and $2,503 by 1987. The administration had implied that this savage increase would be offset by income tax cuts. But the preliminary income tax proposals, just unveiled, would take only $70 off the family's $6,218 tax bill. By contrast, a family with $10,000 in income would have its taxes cut from $446 a year to $16 a year. [January 9, 1978: 92]

According to the editors, "the effects of this egalitarian policy on social tensions are painfully predictable." Indeed, this conflict over "income shares" will be one of the primary dynamics of the postaffluent transition. The concerns are real. The superaffluent aspire to discretion over savings and investments. The affluent expect discretion over spending, either through disposable income or the provision of credit. The subaffluent struggle to maintain discretion over the conditions of basic maintenance. In the meantime the subsistence sector has no discretion at all except perhaps in the unorthodox opportunities of the subculture of the urban underclass.

THE EMPLOYMENT PROSPECT

Within this macroeconomic framework, what is the future of employment policy? Over the last several decades unemployment and welfare have been the principal determinants of federal employment policy. In the last several years the particular condition of youth unemployment has become an additional significant determinant of national policy. In 1978, for instance, the U.S. Department of Labor has budgeted over $2.3 billion for youth programs which were to create almost 1.5 million jobs or training positions. At the same time, however, the number of high school dropouts grew by 80,000 to over 800,000.

But a preoccupation with unemployment figures leads to a neglect of at least four other possible components of a postaffluent employment policy. These are: (1) the insufficiency of earned income, (2) the regional distribution of employment, (3) the quality of work, (4) the emergence of reeducation needs in the adults labor force.

Most of the bulge generation will be absorbed into the work force through service jobs. The emerging service economy which is spatially concentrated in metropolitan areas emphasizes tertiary economic activities—services such as transportation, insurance, finance, management, and real estate. Quaternary and quinary activities, including many governmental activities, education, training, leisure, and cultural activities, and social and medical services, are also important in the postindustrial economy and work force. The skills required of the service worker involve using data, relating with people, and coming up with new knowledge, as opposed to the handling of materials and machines required of the industrial worker in mining, farming, manufacturing, and construction, or what are classified as primary and secondary economic activities.

NEW CATEGORIES OF SERVICE WORK

At the top income levels in the category of service workers are the so-called knowledge workers, people who, according to one writer, are "paid for putting knowledge to work." This is the fast-growing group that includes accountants, engineers, social workers, doctors, nurses, lawyers, computer experts, teachers, researchers, and, especially, managers. Common to all these workers, as opposed to the industrial worker, is their heavy reliance on information, modes of analysis, and the production and distribution of knowledge instead of things. It has been estimated by Daniel Bell (1973) that 25% of the 89 million persons employed in 1975 were knowledge workers; of that group, 15% were professional or technical employees, and 10% were managers, public officials, and proprietors. These are the Type K workers.

Included also within the category of service worker, but lacking the pay- or job-related amenities, are the workers in clerical capacities and counter sales, in the areas included in the service economy. A labor writer has said that "about 80 percent of the employees in the service economy—the bulk of the 'new class'—are fundamentally in the same situation as the old blue-collar worker. The most noticeable difference is that employees in 'the new class' generally earn less" (Tyler, 1973). The reason for lower earnings, he continues, is that the large number of women employed in this sector are generally paid less than men. These lower-paid service workers are known as Type C.

The expansion of new types of clerical jobs, especially in a modern office dependent upon different kinds of data processing, and the expansion of counter jobs where the transaction occurs over a counter, such as in a fast food restaurant or in a government office, are also noteworthy.

A third classification is the Type P worker. This is the type which specializes in protective services. They might range from airport security guards to water testers in an environmental protection agency. In one sense these may also be knowledge workers but the emphasis on surveillance, detection, and control gives this group a different occupational orientation. The knowledge with which they deal is, in many cases, deliberately covert. Also, their relationship with clients is less the provision of a service than the establishment of control.

What is different or useful about this topology? Many writers claim they are a new class distinct from the old classes of capitalists and union workers. Others, deny that notion, seeing a basic rift between the knowledge workers at the top of the scale, and the clerical workers who constitute the bulk of service workers. The separation is in political terms. In occupational terms, however, it could be said that this group is unified in that all workers are rendering services. Indeed, the lower-level workers are rendering services to the knowledge workers in many cases. Whether in the professional or in the clerical levels, all these service workers have in common that they provide services through knowledge-based, social-technological systems in distinction to the capitalists and industrial workers, whose orientation has been to working with machines to produce tangible things as their end product. Data analysis, environmental analysis, and people analysis are the basic tools used. Required for these functions are what might be called a new Three Rs of the service economy—recording (of data), relating (to people), and renewing (of knowledge). Because of the nature of service work, job experiences may become a major component of employment policy.

WORKPLACE REFORM

The evolving maturity of industrial technology and the emerging character of the postindustrial economy has brought about the changes in the nature of jobs and the structure of the workplace described above. Can society, through either public policy or individual action, provide guidance to and direction for these changes? If jobs are plastic to technological forces, can they also be plastic to economic policy and social objectives? If we achieve a better understanding of the centrality of the workplace upon our social satisfactions and stresses, can the objectives of our social development be formulated so as to provide guidance to job design and workplace reform?

Neal Q. Herrick and Michael Maccoby (1974) have suggested four general principles for creating new workplaces that would help the worker improve personally. The principles, and the major requirements needed to assure each one are shown in Table 25.

These admirable principles listed would seem appropriate for many areas of American society, even outside the workplace. Achievement, however, may be much more difficult than articulation. Indeed, the realization of any one of these principles may be in conflict with the attainment of another. The principles of security, equity, individuality, and participation are, however, nonpecuniary components which need to be factored into a job design/job satisfaction framework.

TABLE 25: FOUR PRINCIPLES FOR WORKPLACE REFORM

Principle 1	Security versus the fear and suspiciousness of insecurity
Requirements:	health and safety guaranteed work guaranteed income pension reducing jobs by attrition rather than by firing
Principle 2	Equity versus the envy and resentment of inequity
Requirements:	fair pay differentials profit sharing more responsible work adequately rewarded fair promotions and job assignments
Principle 3	Individualization versus the boredom of being made a cog of the machine
Requirements:	craftsmanship-autonomy continual education opportunity to develop skills and abilities nonbureaucratic treatment of individuals
Principle 4	Democracy versus the passiveness of authoritarian organization
Requirements:	participative management self-management autonomous work groups participation in hiring coworkers representative democracy free speech in the work place

SOURCE: Neil Q. Herrick and Michael Maccoby, "Humanizing Work: A Priority Goal of the 1970's," in Michael Maccoby and Katherine A. Terzi, eds., *Character and Work in America*.

To date, however, it has been easier to plan job development and work reform around issues of technology, productivity, and time. Indeed, recent efforts such as those associated with flexitime have been successful by simply emphasizing a better distribution of entry and exit time from the workplace. Job characteristics do not just happen. Jobs come to be designed for a specific purpose. Tasks can be and are structured and assigned with a specific place in the production process in mind. The interplay of individual goals, alongside the search for increased productivity, has increasingly captured the attention of those who observe and advise management's drive for improved production.

Flexibility in time scheduling and job redesign have implications for the growing awareness of the individual's needs as they affect productivity.

Certain indicators help to establish a policy concern with the quality of work and with the potentials for innovative work reform. Falling or stagnating productivity in the local workplace, rising hourly wages, absenteeism and similar conditions, juxtaposed with the rising educational levels of workers, create the motive force behind the recent renewed attention to work redesign. The goal of these efforts is to develop, if not a creative workplace, at least a tolerated one.

The forty-third American Assembly meeting in late 1973 adopted a pragmatic view: "Improving the place, the organization and the nature of work can lead to better work performance and a better quality of life on the society" (Rosow, 1974: 3).

A more urgent view is taken by Sheppard and Herrick in *Where Have All the Robots Gone?* (1972). In their selected sample of over 1,500 workers, they found that opportunity for personal growth, elements of autonomy, interesting work, and so on were more highly correlated with overall job satisfaction than were good pay or security.

Furthermore, they write:

> Our major emphasis is on the notion that certain kinds of jobs are undesirable, not only for the individual worker employed at them, but, equally important, for the general society.... Our findings lead to the conclusion that for nonauthoritarian workers, jobs with little variety, autonomy, and responsibility can lead to socially and politically undesirable attitudes and behavior.

The future of the nature of jobs is an issue that will not go away any time soon. There are those who can discard this issue by claiming that the increase in leisure time and the concomitant decrease in work time make the issue irrelevant. Since the time spent at the workplace diminishes in relative proportion to the rest of one's time, so does the importance of satisfaction (or dissatisfaction) gained from workplace activity.

Another group will claim that the future of jobs and workplace satisfaction is significant only insofar as pecuniary issues are addressed. People work primarily, if not solely, for cash income (including deferred cash incomes in the form of pensions). In this view the future of the nature of jobs is less important than the future of the nature of income distribution. Problems of workplace satisfaction can be simply addressed by increasing wages or other pecuniary benefits, usually through increases of productivity or through a realignment of the reward structure. Another point of view is that the issue is really political and requires greater democratic control of the corporate economy. This control, some argue, needs to be exercised through better and stronger governmental regulation and planning. Others argue that the control needs to be exercised through more direct worker and consumer control at the place of production. Regardless of the perspective one takes, the nature of the workplace increasingly will become a matter of policy concern.

WORK AND PUBLIC POLICY

The forces of a postaffluent social economy and of a postindustrial technology will bring about changes in the nature of jobs and the character of the workplace. If we assume that work is central and that jobs are plastic, we then must realize that as a society and as individuals we can make some choices and promote some changes.

Before one can usefully begin to structure a study of the relevant techniques and research findings, one needs to explore the possible scope of public policy. What should our workplace goals be as a society? In the social economy of the future, should we even worry about workplace goals?

An interesting perspective is offered in a recent book by James O'Toole (1977). O'Toole does not question the central importance of work in the life of an individual for the social and psychological benefits it holds over and above the economic, in integrating the individual into society as a full participant. What he does question, however, are the commonly held assumptions about work, society, individuals, and economics, which must be broken out of before viable policies can be formulated. These primarily false assumptions, he argues, are constrictions that prevent us from finding alternatives that would bring holistic social and economic benefits. One assumption he discards as false is that "people are unemployed because they lack skills." Policy-making based upon that assumption, which is likely to result in expanding skills training, "merely alters the order of individuals in the job queue" while distracting national attention from the real task of creating jobs; the workers already have the skills or can readily obtain them on the job. O'Toole argues further that these assumptions result in an either-or attitude, a "trade-off mentality," which eliminates from consideration what might be fruitful alternatives. An example is the assumption that cleaning up the environment will necessarily result in higher unemployment, which leads policy-makers to take a do-nothing stance. Instead, he urges, develop job-creating environmental policies, such as a law banning the use of nonreturnable beverage containers, which might create thousands of jobs at the same time as cleaning up the environment. "Future gains in productivity," he writes, "will come from the fuller development and use of the skills, training, intelligence, and education of workers. To more fully tap these human resources will require a new set of assumptions about labor, capital, technology, and productivity." Specific suggestions that he makes are doing away with legal and "credentialist" barriers to employment and designing policies that will create more diversity and flexibility in work conditions and movement into and out of the labor force.

Larry Hirschhorn (1974) discusses the consequences for work and education of the emerging service society and the forces of "antidevelopment" that prevent the emergence of the services' proper postindustrial functions. He differentiates between job growth, the mere expansion in the number of jobs, and job development, "the qualitative changes in the structures of jobs." To avoid stagnation, he argues, public policy-making must concentrate on the development of jobs and not just on an increase in numbers.

Hirschhorn describes the shift in postindustrial society to new sources of productivity, summed up as information and organization, by means of which the service sector becomes the dominant sector in the economy. New needs, possibilities, and skills, however, are contradicted and restricted by attempts to adhere to old structures carrying over from the period of industrialization and modernization. Moreover, different areas within the service sector tend to develop at different rates, causing antagonisms. When such an antagonism occurs, the more advanced service becomes checked, or to use Hirschhorn's term, "rationalized," and effectively downgraded to fit into the older structure. The service is thus underemployed, and society as a whole remains underdeveloped measured against its postindustrial potential.

The crisis in jobs is, to Hirschhorn, an example of an antagonism resulting from unbalanced postindustrial development. A rapidly changing job structure and new education patterns should reinforce each other, leading to dynamic growth in both. But when growth in education outpaces the transformation of the job structure, the old patterns and inflexibility of the job structure cause the development in education to seem "excessive and wasteful." The probable, unfortunate result of this disjunction will be, he suggests, the "rationalization" of education by downgrading it through, for example, reviving more traditional forms of education and vocational training. Such solutions will most likely result in retardation of the potential for development of society as a whole.

What is needed instead, he argues, is the upgrading of the productive "human factor." Reorganization, flexibility, and continually creative changes, those other services involved by it must also change in order for it to develop. He writes: "Information, organization, research, labor mobility, all come to play increasingly important roles in the economy, only when human beings have reorganized their socio-economic activities to bring out the productive potential of these new factors." The alternative, to him, is economic stagnation, further decreases in productivity, and waste.

There are limits to the utilization of educational reform as a device to change the future nature of jobs, yet the educational system does provide policy-makers with a lever that can influence the nature of work. The team that produced *Work in America* (HEW Task Force, 1972) also emphasizes the importance of equivalent forms in school as preparation for smooth functioning of the individual in work; schools up to now have been charged with preparing "industrial man," but changes in society have shifted to a need for preparing the "satisfied worker." They quote Alvin Toffler's description of fitting the schools to the needs of an earlier industrial society:

> The whole idea of assembling masses of students (raw material) to be processed by teachers (workers) in a centrally located school (factory) was a stroke of industrial genius. The whole administrative hierarchy of education, as it grew up, followed the model of industrial bureaucracy.... The inner life of the school thus became an anticipatory mirror, a perfect introduction to industrial society. The most critical features of education today—the regimentation, lack of individualization, the rigid system of seating, grouping, grading, and marking, the authoritarian role of the teacher—are precisely those that made mass public education so effective an instrument of adaption for its place and time. [p. 141]

Times have changed, however, and "industrial man" is no longer a fitting description, since today's workers will no longer accept "authoritarianism, fragmentation, routine, and other aspects of the inherited workplace." The tasks of the schools in the present and for the future is rather, for these writers, to encourage the development of the "satisfied worker." In preparation for today's world of work, today's student needs, and tomorrow's will need, information about the world of work and a realistic appraisal of work, "its meaning, its necessity for life, its rewards, it requirements, and its shortcomings."

To implement this basic shift, they suggest "information-rich alternatives" and "action-rich alternatives." The former include a strong general base in reading, writing, speaking, and arithmetic skills, learning "how to learn," and interpersonal communications skills. Useful also, they write, would be to broaden the curriculum to include courses to learn the procedures of the large work organizations that most students will probably move into, and in entrepreneurship to give students a better basis to "work on their own" should they so choose. Action-rich alternatives include modifying present vocational courses and broadening them into practical courses for all students, in domestic skills, crafts and mechanics, and typing, simulations reproducing work environments; and work-study programs.

A different perspective is offered by Henry Levin (1975). He attempts to anticipate possible reforms that may become instituted in work in the near future, in order to propose "corresponding" reforms in schools to prepare students better for the workplaces they will be entering. Levin describes a "correspondence principle," by which reforms in education are likely to be made when the existing approach and the results it produces do not correspond to changes in the functioning of work organizations. Educational reform thus comes about as a response to the changing societal needs that have arisen. The basic premise underlying this principle is that "schools and other agencies of socialization serve to reproduce the social relations of production. Educational reform, then, becomes functional only when the social relations of production are altered," and not per se.

According to Levin, the "correspondence principle" is not operating at present, and he seeks to reestablish the functional relationship between school and work by proposing a "taxonomy" of educational reforms classified into categories derived from probably coming reforms in work. He suggests that "modifications in work may require workers with different skills, attitudes, and behaviors, and that it is desirable to develop a classification or educational reforms that will be likely to correspond to such changes." Levin's categories of reform are:

- "Microtechnical"—reforms that are "narrowly technical and individualistic in implementation." Examples in work would be a more flexible work schedule and job enlargement. Examples in education do not correspond exactly, but two types of reform in this same category are new subjects and teacher training and retraining.
- "Macrotechnical"—changes in the technical sphere, but wider in scope than the above. Examples of such reforms in work are new staff organization, "revitalization" through educational and training programs, sabbaticals. Examples of such changes in the educational sphere are open classrooms, mastery learning, work-study programs.
- "Micropolitical"—changes in the internal governance of the organization with respect to decisions on production activities and resource allocation." Examples for work are job enrichment and forms of participative management. Examples for schools are changes in governance within the classroom and in the operation of the teaching process, e.g., peer teaching.
- "Macropolitical"—major changes "in the governance and technical conception or organization." Examples of such reforms in work would be worker self-management, employee ownership. Such reforms in schools would be community control, factory-run schools.

As can be seen from the examples above, Levin's reforms in the dual areas of work and education only occasionally correspond on specific points and are not for the most part closely linked. In conclusion, however, he presents a summary of educational reforms "that are consistent with two types of work reforms." Work reforms that result in an increased emphasis on individuality would find their equivalent in educational reforms resulting in (1) educational technology, (2) differentiated staffing, (3) flexible modular scheduling, (4) open schools, (5) educational vouchers, and (6) deschooling. Educational preparation for work reforms instituting increased emphasis on cooperation and participation would result in (1) team teaching, (2) mastery learning, (3) desegregation, (5) micropolitical changes, (5) community control, and (6) factory-run schools. Levin suggests, then, that equivalent forms of organization and decision-making be instituted in schools to prepare the students for the forms in which they will be required to function in the work world.

As American society passes from its spurt of affluence into the postaffluent period, there is a pressing need for the federal government to provide more effectively a policy framework that will insure a full-employment economy and an adequate and rational welfare system. The legislative struggle to provide for this will continue into the next sessions of Congress. Beyond those concerns lies a broader approach to labor-force policy which will emerge in the forthcoming decade. Glickman and Brown (1973), under a grant from the Manpower Administration of the U.S. Department of Labor, suggest that further manpower policy will concern the development of programs to

- Increase the individual's ability to update skills and knowledge in step with technological advance through a mixture of work and study.
- Provide better use of time, space, and facilities through the reallocation of daily, weekly, monthly, and yearly periods of work and nonwork activities. This reduces congestion and overloading at peak periods and underutilization at other times, both in the public and private sectors.
- Offset labor market fluctuations through encouraging or discouraging education and other kinds of nonwork activities in order to accommodate better the supply of labor to variations in demand.
- Increase personnel satisfaction and well-being by providing more opportunities for individual choice of work and other life activities.

Glickman and Brown (1973) go on to claim that it is very difficult to legislate flexibility and suggest that it is government laws and regulations that tend to circumscribe rather than to expand options for the employer and the employee. They suggest that government will have to be challenged to develop new policy instrumentalities that exert positive forces to make flexibility and innovation a working principle in American society and its economy. It might be expected that government as a major employer would itself institute workplace reforms. In the past, however, bureaucratic organization in government has followed, and not led, the evolution of industrial organization forms. Perhaps through revenue sharing particular local governments might be able to experiment more thoroughly, but expectations should remain moderate.

One hopes that the federal and state governments in postaffluent America can encourage and facilitate experimentation in the organization of work. Government should certainly provide for much more research support in this area. It is also likely that federal and state support for expanding public investment in the spheres of continuing educational and recreational facilities will sharply increase, especially in urban areas. The development of urban extension learning activities through American universities has just begun and will continue. Proposals for large-scale public investments in urban recreation have been circulating within the federal government and represent potential support for the evolving leisure ethic.

The worker Alienation Research and Technical Assistance Act introduced by Senators Edward Kennedy, Gaylord Nelson, and fifteen others provides initial support for research, demonstration grants, and technical assistance in these areas; but it has not yet been passed. In Europe, a French law initiated in 1971 is suggestive. This law uses a payroll tax to provide that every major French employer must have at any given time 2% of his work force engaged in full-time study.

Corporate policy will probably be the primary influence on the achievement of a more innovative approach to manpower development in postaffluent America. The real source of innovations must be the private sector. Innovations in workplace reorganization will probably come from the demands and needs arising in particular urban communities. Community values and social norms also contribute to positive or negative job attitudes. In large, depersonalized urban settings, the transformation of oppressive workplaces into more tolerated work environments should be encouraged. In those communities (and labor markets) where there is relatively good access to better leisure opportunities, it is likely that the tolerated workplaces will seek to implement more creative changes. Private businesses should be encouraged to follow and learn from these efforts.

REFERENCES

BELL, D. (1973) The Coming of Post-Industrial Society. New York: Basic Books.
Business Week (1978) "Leveling the middle class," January 9.
GLICKMAN, K. and G. BROWN (1973) Changing Schedules of Work: Patterns and Implications. Silver Springs, MD: American Institute of Research.

HERRICK, N. Q. and M. MACCOBY (1974) "Humanizing work: a priorty goal of the 1970's," pp. 35-47 in M. Maccoby and K. A. Terzi (eds.) Character and Work in America. New York: Andrew W. Mellon Foundation and Harman International Studies.
HEW Task Force (1972) Work in America. Cambridge, MA: MIT Press.
HIRSCHHORN, L. (1974) "Towards a political economy of the service society." Washington, DC: Economic Development Administration.
LEVIN, H. (1975) The Limits of Educational Reform. New York: Basic Books.
LIFTON, R. J. (1975) The Survival of the Self. New York: Simon and Schuster.
MARCUSE, H. (1974) One Dimensional Man. Boston: Beacon Press.
MARIEN, M. (1973) "Daniel Bell and the end of normal science." Futurist (December): 10-17.
MUMFORD, L. (1938) The Culture of Cities. New York: Harcourt Brace Jovanovich.
OPHALS, W. (1974) "The scarcity society." Harpers (April): 45-49.
O'TOOLE, J. (1977) Work, Learning and the American Future. San Francisco: Jossey-Bass.
ROSOW, J. (1974) The Changing Nature of Work. Englewood Cliffs, NJ: Prentice-Hall.
SCHON, D. (1973) "Technology and social change," pp. 86-108 in Y. Hayashi (ed.) Perspectives on Post Industrial Society. Tokyo: University of Tokyo Press.
SHEPPARD, H. L. and N. HERRICK (1972) Where Have All the Robots Gone? New York: Free Press.
TYLER, G. (1973) "A labor view of the new class." Amer. Federationist (October): 10.

10

The Diminishing Urban Promise: Economic Retrenchment, Social Policy, and Race

HAROLD M. ROSE

☐ BLACK MOVEMENT TO THE CITY, beginning with the Great Migration (1915-1919), held promises of economic progress that was not possible within the prevailing racial caste system which characterized the South. The development of an effective interpersonal communications network fueled continued movement between the South and the North for more than two generations. During that period the South experienced a net loss of more than five million blacks to selected cities of the North and West. The magnitude of this flow began to show signs of diminution during the last half of the sixties. During the first half of the seventies there was evidence that the tide had finally turned and black migrants from the North were moving South in larger or at least equal numbers to those who were moving in the traditional migratory channels. The pattern of population change which spans the period 1910-1975 coincides well with the fortunes of regional economic development, wherein the less developed South was a region of surplus population that provided unskilled labor to the more developed North. By 1975 the developmental differences that distinguished these two regions had begun to subside. Now it was the South that represented the growth pole and the North with a surplus of black labor which tended to be concentrated in the nation's largest and older manufacturing cities. For that segment of the black population which viewed Northern migration as an adaptive mechanism that would allow them to upgrade their economic status it appeared that this was rapidly becoming a less viable option.

Even though regional differences in level of economic development have been minimized, the national rate of economic growth has been severely retarded during most of the seventies. This slowdown in economic growth has made it necessary to investigate a host of policy options designed to minimize that set of dual problems which have accompanied the economic downturn, inflation and unemployment. These twin problems are perceived differently by population groups as a function of their position in the economy. Polls show that most Americans currently list inflation as the number one problem confronting the American people, while those organizations representing blacks generally tend to emphasize the unduly high unemployment levels that are pervasive in the nation's larger black communities. It should be acknowledged, however, that blacks like other ethnic groups in American society are found to span the spectrum of economic well-being although a smaller percentage have been able to escape low-income status. Improvements in the black social class structure since World War II have basically been associated with movement out of rural occupations and other low-status jobs that were traditionally open to blacks in the southern economy and into jobs in manufacturing and middle-level services. These changes have led to an expanding black middle class, which has prompted some social commentators to accuse civil rights organizations of failure to publicize this recent group progress.

Employing a variety of indicators of economic progress, such as median income shown in Table 26, it is clear that the mean position of blacks has improved. However, the degree of income inequality among blacks has increased. Therefore, it is apparent that generalizations regarding blacks have become less meaningful. The primary issue which this paper will address is whether the gains acquired by blacks during the previous decade can be maintained in view of the shifts in national economic well-being. Table 26 shows that there has been very little reduction in the ratio of black to white income during the 1970s, a period of slow economic growth. Thus in a period of economic retrenchment what can we expect in the way of social policy options to protect the economic gains of one generation and to reduce the clustering of a disproportionate segment of the black population toward the low-income end of the income spectrum.

TABLE 26: MEDIAN INCOME OF FAMILIES: 1950 TO 1976

	Race of Head		
Year	Black and Other Races	White	Ratio: Black and Other Races to White
1950	$1,869	$ 3,445	0.54
1951	2,032	3,859	0.53
1952	2,338	4,114	0.57
1953	2,461	4,392	0.56
1954	2,410	4,339	0.56
1955	2,549	4,605	0.55
1956	2,628	4,993	0.53
1957	2,764	5,166	0.54
1958	2.711	5,300	0.51
1959	3,161	5,893	0.54
1960	3,233	5,835	0.55
1961	3,191	5,981	0.53
1962	3,330	6,237	0.53
1963	3,465	6,548	0.53
1964	3,839	6,858	0.56
1965	3,994	7,251	0.55
1966	4,674	7,792	0.60
1967	5,094	8,234	0.62
1968	5,590	8.937	0.63
1969	6,191	9,794	0.63
1970	6,516	10,236	0.64
1971	6,714	10,672	0.63
1972	7,106	11,549	0.62
1973	7,596	11,549	0.62
1974	8,265	12,595	0.60
1975	9,321	14,268	0.65
1976	9,821	15,537	0.63

SOURCES: U.S. Bureau of the Census, Current Population Reports, Special Studies, Series P-23, No. 54, The Social and Economic Status of the Black Population in the United States, 1974, p. 25; and U.S. Bureau of the Census, "Money Income and Poverty Status of Families and Persons in the United States: 1976, "Current Population Reports, Series P-60, No. 107, Table 2, p. 9.

SLOWED ECONOMIC GROWTH AND ITS CONSEQUENCES FOR THE BLACK MINORITY

The American economy of the seventies is partially a product of a major transformation of the nation's industrial structure since World

War II. The most significant change to take place in the post-World War II economy has been the relative decline in the number of manufacturing jobs and the growth in service jobs. During the period 1960-1975, both the Middle Atlantic and East North Central census divisions experienced an absolute decline in manufacturing jobs (Sternlieb and Hughes, 1977: 230). Table 27 illustrates that there were salient employment trends. These two census divisions constitute the core of the American manufacturing belt. While manufacturing activity is declining nationally, the South has been a major recipient of new jobs as manufacturing has shifted out of the nation's industrial core. By 1975, the South possessed a slightly larger number of manufacturing workers than did the Northeast. The interregional movement of manufacturing jobs from the North to the South has been described as a form of industrial filtering by Thompson. It is said that older industries which

TABLE 27: PERCENTAGE CHANGE IN NONAGRICULTURAL EMPLOYMENT

Area	1965-1975	1970-1975
North Total	14.767	1.845
a. Major metropolitan regions	10.861	-0.778
b. Smaller regions	20.145	5.384
South Total	42.724	16.692
a. Major metropolitan regions	56.580	15.034
b. Smaller regions	37.101	17.476

PERCENTAGE CHANGE IN MANUFACTURING EMPLOYMENT

Area	1965-1975	1970-1975
North Total	-10.258	-11.287
a. Major metropolitan regions	-14.075	-14.363
b. Smaller regions	5.615	-7.612
South Total	21.716	1.582
a. Major metropolitan regions	24.557	-3.427
b. Smaller regions	20.759	3.446

NOTE: For the purposes of the table the North is defined as Maine, Vermont, New Hampshire, Connecticut, Massachusetts, Rhode Island, New Jersey, New York, Pennsylvania, Ohio, Indiana, Illinois, Michigan, Wisconsin, Minnesota and Iowa. The South is composed of Virginia, West Virginia, N. Carolina, S. Carolina, Georgia, Florida, Kentucky, Tennessee, Alabama, Mississippi, Arkansas, Louisiana, Oklahoma, and Texas.
SOURCE: Department of Labor, Bureau of Labor Statistics, Employment and Earnings by States and Areas, 1977.

originated in high-order urban centers eventually filter down to low-order centers when skilled labor is no longer necessary for their operation (Thompson, 1977: 110-111). He further states in regards to this shift that the South "is going through the industrial and post-industrial age at the same time" (p. 188). The slowdown in the formation of manufacturing jobs in the nation's industrial core has eroded the hopes of limited skill black migrants who were dependent upon these jobs to bring about improvements in their economic lot and that of their progeny.

DIFFERENTIAL ACCESS IN REGIONAL JOB MARKETS

In the past Southern-born black migrants were shown to have entered Northern labor markets at a relatively low level in the occupational structure. Nevertheless, they have been shown to have both higher earnings and higher labor force participation rates than their Northern-born black counterparts. Yet black migrants to the North lag approximately twelve years behind their northern-born counterparts in integration into the majority stratification system (Hogan and Featherman, 1977: 104). In other words, Northern-born blacks are more likely to have jobs similar to whites than are recent migrants. This possibly reflects the nature of the structure of employment opportunity available to first generation northern migrants because of the negative value employers often assign to attributes of southernness. The slowdown in migration to the North could possibly further delay black participation in the majority group occupational stratification system. If, however, there is a corresponding opening up of Southern labor markets not previously accessible to blacks, enthusiasm for interregional migration on the part of blacks of Southern residence might be dampened. There is some evidence which indicates that young black entrants to the Southern labor force have reduced the gap in income between themselves and same age white cohorts. On the contrary the gap between young blacks and white males has gradually widened in the North Central region and the West since 1970 (Smith and Welch, 1978: 7). One reason for the increased racial disparity is that the decline in Northern manufacturing jobs has been particularly harmful to blacks. It is questionable though, if Southern occupational progress has been associated with manufacturing shifts, since much of the redistribution of manufacturing activity has been located in Southern nonmetropolitan counties where few blacks reside. If this is true, the inference is that young blacks are enjoying some success in penetrating the postindustrial (largely service) sector of the Southern economy. But the plight of the potential migrant with limited skills, who in the past has chosen to settle in the American manufacturing belt, might have been aggravated by the previously described employment shift. There is evidence which suggests that migrants from the most severely deprived Southern nonmetropolitan counties are continuing to migrate out of the region.

EVIDENCE OF BLACK PROGRESS IN OUR EXPANDING JOB MARKET

Improvements in educational quality and years of education completed by blacks have led to a significant increase in the size of the black white-collar labor force. But this change could not have occurred without a corresponding relaxation of discriminatory policies which operated previously to confine blacks to a narrow range of jobs. Freeman (1978: 55-57) indicates that since 1964, both black males and black females have made good progress in penetrating the market for upper-level white-collar jobs which were not previously open to them. Male progress, however, was shown to lag behind that of females. Some writers contend that these changes are in large measure associated with the monitoring of employment practices by federal agencies that came into existence to support Title VII of the Civil Rights Act of 1974. Smith and Welch recently challenged the findings of some previous research which attributed the closing of the black-white income gap to the role of affirmative action programs. They find little evidence in support of the notion that these changes are primarily related to government efforts to reduce discrimination (Smith and Welch, 1978: 24 and 47-50). It is clear that this is likely to remain a debated topic for some time to come. But regardless of outcome, blacks have been able to acquire access to job markets which were previously closed. Penetration of these markets has had a substantial impact on closing the ratio of black to white earnings. However, as with income, employment opportunities have been spread unevenly among the black population. During the period in which most employment progress was being recorded through the removal or at least the reduction of racial barriers to employment, evidence regarding labor force participation rates suggests that many blacks have simultaneously been withdrawing from the labor force.

THE IRREGULAR ECONOMY AND DECLINING
BLACK LABOR FORCE PARTICIPATION RATES

Since World War II, black labor force participation rates have declined by almost 17% for black males, while showing an increase of almost 5% for black females. The magnitude of the rate of decline for males has been greatest among the extremely young and the older segments of the work force, although a moderate decline was detected throughout the age structure. Among males 20-24 and 25-34 there was a downward shift of approximately 7 and 5% respectively. There has been an increase in labor force participation rates among women in general, but most of this increase is associated with the increasing frequency with which white women have entered the labor force during this period. The largest increase in level of participation among black women has been in the 25-34 age group, where an almost 15% increase has occurred. The major increase in this age group appears to be associated with an increase in opportunity in nontraditional jobs. The timing of the increase coincides with the expansion of opportunity in the job market, first becoming noticeable in 1967. The sharpest changes in male labor force participation rates have occurred in the seventies. Farley (1977: 195) notes that in 1975 three-quarters of the adult white men were in the labor force, whereas only two-thirds of the adult nonwhite men had or were seeking jobs. The obvious inference is that the increasing gap in labor force participation rates between the races is in part related to the higher percentage of discouraged workers among blacks.

The question of what role this growing segment of the nonworkforce population plays in the social economy itself is one of some importance, but is seldom investigated. A significant share of such persons are in school and have thus simply deferred entry into the labor force. However, another segment of this population draws its sustenance from the irregular economy, from both its licit and illicit components. It is this segment of the economy that we know least about. A classification scheme that may be useful in developing an understanding of the entire economy distinguishes between the regular and irregular economies. The regular economy includes the primary and secondary labor markets. The irregular economy includes jobs in the secondary market as well. Fusfeld in describing the irregular economy states that jobs in the irregular economy include activities that are basically developed to satisfy the needs (good and bad) of the black community.

He further indicates that the informal structure of the irregular economy promotes work habits that are inimical to employment in the more highly structured regular economy (Fusfeld, 1973: 34-36).

Friedlander finds that the illicit sector of the irregular economy holds a greater attraction for many segments of the population than do prospects of employment in the low-wage secondary labor market. More specifically, he states that "the illegal market of criminal activities is readily available and serves as an attractive alternative to legitimate employment available to a slum dweller, offering high renumeration and status, flexible hours, glamorous activities and excitement" (Friedlander, 1972: 113). Rising crime rates in the black community are thought to be partially associated with the operation of the illicit component of the irregular economy. Participation in the irregular economy is often a high-risk venture that allows one to satisfy only one's ephemeral needs. But for those who fail to complete high school during a period of slow economic growth, the irregular economy may appear to represent one of only a few viable options.

In the preindustrial rural and small town South dropping out of school was often prompted by the need to contribute to the support of the family in an economy that could readily absorb persons with limited skills. While pay was poor, these were at least socially sanctioned roles. In today's postindustrial service economy, with its emphasis on technical skills, high school dropouts can increasingly look forward to only limited earning power. The psychological depression which often accompanies the status of persons who are unable to penetrate the primary sector of the labor market often pushes them in the direction of the irregular economy. Thus persons entering the labor force for the first time and not possessing that set of traits demanded by the primary sector of the labor force may choose to seek their sustenance from the irregular economy rather than the secondary economy. At this point there is limited evidence indicating who these participants are in terms of migrant status; however, there is some evidence which indicates that Southern-born black migrants while having lower unemployment rates than their Northern peers, are more often found outside of the labor force in the host environment. This is not to say that black migrants have a greater affinity for being attracted to the irregular economy that do nonmigrants. In fact, there is some evidence which would suggest a lower propensity for participation among such groups than persons socialized in environments where elements of the irregular economy are more prevalent.

WIDENING OPPORTUNITY SCREENS AND
DIFFERENTIAL BLACK ACCESS

The decade of the sixties saw blacks make significant progress toward the achievement of equality of opportunity. The latter part of that decade, even though characterized by chaos, was a period in which the screens of opportunity were widened, and there were blacks available to capitalize on the changing availability of opportunity. The period of most rapid economic gain occurred during a period of aggregate economic recovery and growth.

But the early years of the seventies saw the nation once again enter a period of economic stagnation. The consequences of this down turn could have the effect of dampening the progress initiated during the previous decade. For that segment of the black population who benefited from an expansion of the screens of opportunity, their economic future now appears less secure. Even more tenuous is the projected circumstances of those whose marketable skills did not permit them to participate in the expanding opportunity structure. The latter segment of the black population, because of its inability to readily alter its status, has been the topic of a number of recent economic and social analyses.

Some social commentators have identified the segment of the population lacking marketable skill as an underclass whose social class position is not likely to show much improvement even during periods of rapid economic expansion. Wilson has defined the underclass, though, as that segment of the population which falls below the official designation of low income (1978: 181). His definition includes approximately one-third of the black population in 1974. Thus a sizeable segment of the black population is still outside of the economic mainstream. A number of arguments have been fostered to explain the economy's inability to absorb this segment of the population. Some analysts contend that this segment of the population is hampered by a set of negatively valued traits. Freeman speaks of "the burden of background" as the more important variable slowing the economic progress of black males at present (1978: 61-64). This argument attributes to declining importance the role of racial discrimination and indicates that the major deterrent to continued economic progress is basically related to social origins or family background. Such arguments are often employed to alleviate the social order of responsibility for the plight of those who have not escaped a precarious existence and thus minimize the need for public policy interventions. Emphasis here, in terms of the position of blacks in the economy, has essentially assumed a two-class system: one encompassing the middle class and another the lower class. It is obvious, however, that such a system fails to accurately describe the occupational structure of the population. A sizeable segment of the migrant population who gained a foothold in the industrial economy should be classified as working class. The question then is: Will a shift from an industrial economy to a postindustrial economy delay or abate the economic progress of the offspring of the black working class? At the moment the black sons and daughters whose parents are forty to fifty-four years of age are at greatest risk of having the dream of economic security deferred because of their social origins (nonmiddle class status) and position in the economy (the manufacturing sector). The sons and daughters of the nonblack working class of a previous generation constitute a sizeable segment of today's middle class.

RACIAL COMPETITION FOR SCARCE RESOURCES IN LAGGING SECTORS OF THE ECONOMY

The most recent black gains took place during a period of economic expansion. During the downturn years that characterized the first half of this decade there was a slowdown in the general rate at which blacks were able to close the income gap that separated them from their white counterparts. Changes in the rate of economic growth led some to fear that the gains blacks had experienced during the previous decade might be lost as a result of the pronounced economic reversal. A recent assessment of this situation by Farley (1977: 203-206) indicates that the gains of the sixties were not lost during the early seventies. But he is careful to point out that the income disparity between black and white males was still high, even though much of the gap separating black and white females had been eliminated. He as well as others indicate that a further increase in levels of median family income has been made difficult by the sharp rise in low-income black female-headed households. The aggregate improvement of the economic position of blacks, however, often fails to demonstrate the growing income inequality within the black population. Villemez and Wiswell (1978: 1019-1034) demonstrate the impact that aggregate economic gains have had on the internal distribution of income. They indicate that there has been an increase in the degree of income inequality among blacks. This trend has become most pronounced in North Central and Middle Atlantic census divisions, but has been somewhat ameliorated by the availability of welfare payments. This leads to the conclusion that any assessment of black economic progress is likely to be misleading if aggregate racial differentials alone are employed in an assessment of black economic progress. Likewise, the further narrowing of the income disparity between blacks and whites is in large measure dependent upon a reduction in the intensity of interracial competition for resources in lagging sectors of the economy.

The relative decline in the manufacturing employment creates additional pressures on the economy to absorb blacks into those economic sectors that provide both security and a decent wage. Manufacturing has been the hardest hit by the most recent recession which got underway during the early years of this decade. The level of unemployment among males reached its peak during 1975 before beginning to show signs of recovery, but most speculations on the nature of "postindustrial society" suggest that manufacturing will not be a rapidly expanding area. It is generally agreed that male unemployment losses are basically associated with those sectors of the economy that are most sensitive to cyclical changes (Henry, 1978: 58). These areas have been identified as the durable goods, transport, and construction sectors of the economy. The heavy dependence of blacks on the durable goods sector was noted by Brimmer (1972: 40). In the second quarter of 1975 approximately 35% of both black and white unemployed workers were associated with the manufacturing sector. In the previous year about 28% of both groups were employed in manufacturing. The level of concentration of their unemployment, however, differed in the two primary sectors of durable goods manufacturing—primary metals and automobile production. Approximately 28% of all black workers in manufacturing were in these two sectors, while 18% of all white manufacturing workers were in these sectors (National Urban League Research Department, 1976: 9). Since the Northeast and the North Central division represent zones in which manufacturing activity is on the decline and since it is in these geographic divisions that black and white workers with limited skill tend to be concentrated in this sector of the economy, it is expected that competition for limited jobs in the Northern industrial belt will become more intense.

THE ROLE OF UNIONS IN PROVIDING WHITE ADVANTAGE

The manufacturing sector layoffs tend to be concentrated more often in the less skilled sectors. Operatives were about twice as likely to be unemployed than craftsmen. Blacks tend to be disproportionately represented among the former. The historic exclusion of blacks from the skilled trades has had the effect of denying both greater income and job security. Although it is generally agreed that the industrial unions have treated blacks more favorably than have trade unions, even among the former there has been evidence of widespread discrimination and denial of opportunity to hold key positions in the union hierarchy. For instance, the Bethlehem Steel Corporation until recently maintained job classifications open to whites only. These practices prevailed with union sanction. Only after an executive order directing the firm to eliminate such practices was issued on January 15, 1973 was the policy terminated (Foner, 1974: 427). Dual seniority systems and closed job classifications which met with the approval of both unions and firms have assisted in keeping blacks confined to those jobs that were the least desirable. The elimination of such advantages favoring white blue-collar workers can be expected to heighten previously existing antagonisms. While working-class consciousness is thought to prevail among manufacturing workers, that consciousness is weakened by racial antagonisms. One investigation of this issue found that "on the whole, race is the most important influence on class consciousness, although nationality differences help to account for variations in attitudes within the mainstream working class" (Leggett, 1968: 117). While unions have assisted workers in securing gains from management, their role in providing such advantages has not been racially neutral.

In Detroit, black unionists were described as exhibiting a militant posture prior to the onset of the period of black political militancy in the middle and late sixties. Blacks sought more power within the union hierarchy as a means of improving working conditions in the plants and to improve the status of black workers. As late as 1968, fewer than 3% of the employees in the skilled trades were black (Foner, 1974: 411). Militancy among black unionists was heightened during the late sixties. There are indications that increased militancy among black auto workers was associated with the hiring of young black workers following an upswing in automobile demand. The hiring of additional black workers was also an attempt by industry to reduce the size of the hard-core unemployed following the Detroit riot in 1967. This new cadre of workers expressed dissatisfaction with working conditions that had gone unchallenged by older workers. Finding the union grievance procedure unresponsive to their concerns they organized to develop a set of tactics that would push both unions and management to take action on their concerns.

The effectiveness of black militant workers was partially strengthened by black worker isolation in selected units of the firms, as blacks constituted 60 to 75% of the work force at some plants (Foner, 1974: 411). The Revolutionary Action Movement was formed at the Dodge Main plant in the late 1960s and spread to approximately twelve other plant sites in a short period of time. For several years the struggle between militant black workers and both union and management continued. It is thought that internal dissension growing out of conflicting goals and the firing of local organizational leaders led to the dissipation of the movement. By 1971 it was said that "the League of Revolutionary Black Workers had become history" (Georgakas and Surkin, 1975: 164). The support of the league was diminished with the layoff of auto workers during the slump of 1970. As the young militant workers represented the most recently hired, they were the first to be fired. Foner (1974: 422) indicates that as early as 1970, most black workers "were marching to a different drummer." Older workers who were less militant, however, readily acknowledged that black union representation had increased as a result of league action.

In both the industrial unions and the craft unions, blacks have been blocked in their attempt to seek parity with white workers in the same firm. Title VII of the Civil Rights Act of 1964 made it potentially possible to correct some of the most blatant grievances. One of the more important gains associated with the passage of this legislation has been the opening of apprenticeship programs. By 1976 blacks constituted 9.4% of all persons in registered apprenticeship programs. They did, however, constitute almost 15% of the apprentices in the transportation equipment sector of manufacturing. It should be noted though that 52% of all registered apprentices are in the construction industry and that there the share of black apprentices basically conform to the national average. In the generation that has passed since 1947 no fewer than 125,000 persons have been in apprenticeship training programs each year, with the number exceeding 200,000 per year in almost half of the years. Yet only in the most recent period have blacks been able to penetrate this elite blue-collar work force. Black entry into the skilled trades in the manufacturing sector has come at an inopportune time, since manufacturing is becoming a less important sector. Likewise, the industrial Northeast and North Central division lost more than one million manufacturing jobs in the ten cities that were major target of black migration during the period 1947-1972 (Sacks, 1977: 139-140). This decrease in manufacturing jobs and the elimination of the double queue can be expected to intensify competition between the races for jobs in sectors of the manufacturing economy.

THE STRENGTH OF PUBLIC SECTOR EMPLOYMENT

The decline in manufacturing jobs has frequently led unskilled blacks to become increasingly dependent on public sector jobs. Public sector service employment has been called upon to provide opportunities for the unskilled worker as manufacturing jobs relocate to the suburbs or are disappearing totally and also to provide higher-skill opportunities for blacks entering the middle class. Since 1962 public sector employment has grown somewhat faster than jobs in the private sector although public sector employment accounts for only a small percentage of the total labor force. Blacks exhibit a greater dependence upon public sector employment than do whites. Approximately 16% of the federal civilian labor force was black in 1973 (Brimmer, 1974: 100). The federal level of employment is thought to exceed that at both the state and local level. It is unlikely, however, that the public sector will be able to provide an adequate refuge for displaced blacks and other unskilled workers during a period of economic retrenchment.

As the black population in the larger cities increases, it will likewise increase its share of public sector jobs. This latter prospect is predicated on the notion that an increasing share of social services will be targeted for the black population. If such is the case, there is likely to be support for additional black providers. In those cities where blacks have gained a modicum of political power, effort is being exerted to open up a greater number of jobs to blacks in local government. In Atlanta, blacks now constitute more than half of all persons employed by the city government (Jones, 1978: 116). Efforts of the mayor to increase the number of blacks in certain city jobs have antagonized elements of the white community. In those cities with black mayors who insist on increasing the number of black policemen, conflict between the mayor and representatives of white police is quickly heightened. Both Mayor Jackson of Atlanta and Mayor Young of Detroit have incurred the wrath of their local police by insisting on a greater number of black employees and speedy promotions for black personnel. Even though the size of Atlanta's black work force has increased since the election of Mayor Jackson, most blacks are still concentrated in the laborer-occupational category, with some evidence of increase in black personnel in the skilled worker categories. The political struggle for improved jobs shares in police and fire departments in large cities where blacks are on the brink of becoming the majority population often leads to racial conflict, with the courts frequently being asked to intervene.

During the recovery period from the most recent recession, public sector employment has shown less ability to bounce back than the private sector, although the jobless rate is lower for individuals previously employed by government than in the private sector. Added pressures are mounting for a curb of growth in public sector employment. Some believe these pressures, given the fiscal dilemma cities find themselves in, will result in policies designed to place ceilings on the public sector work force and, in some instances, to actually reduce it (Bahl et al., 1978, p. 17). In the most severely distressed large cities, public sector jobs were observed to decline between 1972 and 1976. In those large cities in which public sector jobs declined most, blacks constituted a sizeable share of the labor force. A further increase in the level of economic distress in these cities will doubtless lead to an elimination of those jobs which can be removed with the greatest ease. Laborers in big-city sanitation departments are probably the most vulnerable in terms of union protection. It is precisely in this segment of the public sector work force that blacks are most likely to be found. The bargaining power of this group is less substantial than that of the more pivotal groups such as the fire and police departments, partly because sanitation workers are held in lower esteem than police or fire personnel. The growing militancy of public service unions, however, is destined to accelerate as taxpayers push for a reduction in the public sector work force. The most vulnerable segment of the public service provider group are those areas in which unskilled blacks constitute the largest share of the work force.

The vulnerability of the black work force during a period of economic retrenchment is not confined to limited-skill blue-collar and service workers although the vulnerability of higher-skilled groups is likely to be much smaller. Black professional workers continue to be concentrated in a narrow band of professional job categories dominated by teaching, health services, and social and recreational workers. Public elementary school teachers represent the single largest professional job category occupied by blacks. There was an approximate 33% increase in this category among black workers during the sixties. Much of this growth can be attributed to the hiring of black teachers in non Southern markets in response to the rapid increase of black children in Northern cities during that period. Prior to World War II most black teachers were situated in the dual school systems of the South. During the 1950s Southern segregationists frequently used the argument that they provided professional opportunities for blacks in a dual school system, while Northerners were advocating school desegregation and denying blacks the opportunity for employment in Northern school districts. In 1967 data showed that most black teachers in Northern school systems were employed in school settings where most of the students were black (Farley and Taeuber, 1974: 902). The index of segregation for black teachers and black students was extremely low in a large number of American cities and extremely high for black teachers and white students in these same cities. It appears that an informal quasi-double queue was at work in the assignment of black teachers. Thus the growth in the number of black teachers in Northern school districts is positively associated with the growth in the number of black students.

The double queue becomes more apparent in those school systems where black students have already become the majority. In 1966 the District of Columbia school system had an elementary school enrollment which was 91% black and at the same time its elementary school teachers were 83% black (Silver, 1973: 32-34). On the other hand, majority white suburban school districts seldom employ black teachers unless there is a significant increase in black enrollment in the district. Washington represents an extreme case in levels of black teacher concentration among the schools studied by Farley and Taeuber. Only in Richmond and Birmingham did such extreme vestige of a dual school system exist as late as 1967. The most extreme isolation of black teachers in Northern school districts at this date was to be found in St. Louis, Cleveland, Dayton, Milwaukee, Detroit, Toledo, and Indianapolis. During the period from 1964-1967, the Milwaukee public schools hired an additional eighty black elementary teachers and 78% of these were assigned to schools that were more than 80% black. Such assignment practices resulted in black teachers having only limited experience with white students, and white students seldom having blacks as teachers. Of the 285 white teachers hired during this same period, a net of only eight additional white teachers were assigned to the predominantly black schools. These patterns provide additional support for the theory of a double queue in the public schools of large American cities.

Since most black teachers in Northern school districts are likely to have less job tenure than their white colleagues, they too are more vulnerable to job layoffs during periods of economic retrenchment. Declining enrollments and taxpayer dissatisfaction with municipal budgets could lead to reduction in school teaching staffs. While white teachers are subject to the same risks, they have greater mobility in securing teaching assignments in systems outside of large central cities. Black teachers, however, may be provided additional security resulting from the lessened willingness of white teachers to accept assignments in predominantly black schools.

THE CHALLENGE TO THE LEGITIMACY OF CORRECTIVE SOCIAL POLICY

During much of this century civil rights organizations have worked assiduously to secure legislation and/or judicial decisions which would assist in altering the status of the nation's black population. State and federal action in the latter years of the nineteenth century did much to place blacks in a semipermanent role of second-class citizen. The official sanctioning of group status by governmental bodies at all levels made it extremely difficult for most blacks to escape from poverty or at least a marginal economic existence. In order to reduce the disadvantages which accompanied previous public action, it was necessary to provide substitute action designed to overcome a prior set of negative sanctions. During the first two generations of this century, the gains associated with attempts to alleviate status disadvantages were sporadic and limited in scope; although they did serve as building blocks for further gains. Major breakthroughs associated with these efforts did not take place until the nineteen-fifties and -sixties. The most notable decisions relative to the removal of discriminatory barriers have occurred in the fields of discrimination in education, housing, employment, and voting rights. The public action of the late sixties, which was designed to guarantee rights, is viewed by some analysts as moving the struggle for black equality from the arena of protest to the arena of politics. Preston, in reviewing this situation, concludes that the problem of implementing policies designed to guarantee rights is likely to be extremely difficult. He further states "that the recession of the 1970s led to retrenchment, and that many previous gains have either been wiped out or have become subject to question" (Preston, 1977: 246). His assessment of the situation indicates that a new white attitude is emerging which supports the notion that blacks have gained too much. It is not by coincidence that the challenge to the legitimacy has accelerated at the same time as the urban promise is diminishing.

The most obvious resistance to policies designed to facilitate black entry into the economic mainstream has been associated with attempts to desegregate the schools and to operationalize affirmative action policies. The issues of busing and white flight, both reactions to school desegregation efforts, have led the Congress and the Court to initiate tactics designed to reduce the scale on which corrective action is taken; for example, Congress has attempted to prohibit the Department of Health, Education, and Welfare from approving desegregation plans where busing would be involved. Likewise, the Supreme Court's position on school desegregation plans appears to be undergoing change. Its recent dictum on segregative intent and interpretations of the principle that the scope of the remedy fit the scope of the violation (at least in Northern school desegregation cases) illustrate the caution that it is beginning to exhibit as it proceeds to refine its interpretation of how rights should be guaranteed. Reactions of both the general public and government officials serve as indicators that the perceived gains of a few years ago may be eroding, as apparent fears of white resource loss grow in scope.

The various branches of government which were previously active in removing barriers to black entry into the American mainstream are being urged by their constituents to go slow. Some of the strongest opposition, in terms of attempts to influence government response, has been in the area of affirmative action policy. Unlike policies designed to promote school desegregation which can be invalidated by changing residence or establishing alternative educational systems, affirmative action policies are not as easily circumvented without obvious attempts at collusion. This fact has led to counteractions designed to invalidate the legal bases upon which such policies are founded. Affirmative action policies, however, are viewed by some as an example of reverse discrimination. The now famous Bakke case has been described by a number of social scientists as a reverse discrimination suit.

There is growing evidence that the race-related policies of the last twenty years which were designed to assist blacks and other minorities are currently being challenged by whites on the ground that whites are being denied their basic rights. The employment of numerical goals or quotas represents the target of much white opposition. Such goals have often been used to ascertain if employers or institutions of higher education were making good-faith efforts at providing opportunities to members of minority communities. It was expected that the Bakke suit would clarify once and for all the constitutional bases of affirmative action policies. The Supreme Court rendered its decision in a five-four vote in July 1978. The outcome of that decision was to acknowledge that Bakke had been unconstitutionally denied admission to the medical school of the University of California at Davis as a result of the positions set aside for disadvantaged students. In addition, the Court stated that, while quotas are unallowable, race can be employed in admissions decisions. The question of the employment of race as a factor in satisfying affirmative action goals is considered unclear by many persons and institutions. The mixed reaction to the Court decision largely reflects one's philosophical predilection.

Black and other minorities view the decision as a setback for minority populations. It appears that this position was supported by Supreme Court Justice Thurgood Marshall who said, "I agree with the judgement of the Court only insofar as it permits a university to consider the race of an applicant in making admissions decisions (Chronicle of Higher Education). He further stated, "I fear that we have come full circle." The latter statement refers to his perception of the manner in which the Court has moved from Plessy vs. Ferguson in 1896 to Bakke vs. the Regents of the University of California in 1978. Others saw this decision as providing continuing support for affirmative action, and simply against the existence of racial quotas. There are also those who contend that there no longer exists a set of clear guidelines specifying how affirmative action goals might be satisfied. This lack of clarity was recently reflected in the legal suit filed by Sears, Roebuck and Company against the Department of Health, Education and Welfare charging that conflicting demands made it impossible to comply with affirmative action guidelines. A number of black spokesmen, however, view the Sears action as an attempt to drive the final nail in the coffin of affirmative action. If this is true, both the public and private sectors of the economy seem to be engaging in tactics which will further weaken an already weak affirmative action program, at least in its ability to generate results. It is expected that the Supreme Court will hear other affirmative cases on its new calendar and will not continue to hedge on the constitutional merits of affirmative action.

Although the constitutional question remains unresolved, it is clear that affirmative action generates antagonisms between blacks and individuals who view such programs as denying opportunity to whites who must compete with blacks and other minorities for scarce resources. The loudest opposition on this issue has frequently come from representatives of southern and eastern European ethnic groups. Organizations representing selected members of these ethnic bodies were often among those filing briefs as friends of the court in support of Allan Bakke. Likewise, white ethnic scholars have been in the forefront of the fight to demonstrate that affirmative action programs are often tantamount to reverse discrimination.

Since institutions of higher education are encouraged by affirmative action policy to expand employment opportunity to minorities and women, this could be expected to lead to weakened support by white males in a context of declining opportunity. In a slow-growth economy, the struggle for opportunity approaches a zero-sum game. Opposition by white males under these circumstances is thought to be related to the threat of loss for themselves or the prospective status of their progeny. In support of this position, Turner and Singleton (1978: 1012) in addressing themselves to the fear of minority penetration of the academic market said: "However, in recent years, academics have been less vocal as affirmative action policies in a tight job market have seriously threatened, for the first time in history, their economic positions."

THE WHITE ETHNIC POSITION ON AFFIRMATIVE ACTION POLICY

Ethnic scholars of southern and eastern European origin, whose foothold in academia is of more recent origin than their white Anglo-Saxon counterparts, might be made even more anxious than the average white academic regarding the impact of affirmative action policies on their status security. Greeley is of the opinion that access to upper-level occupational opportunities appear to be blocked to Catholics with higher education. More specifically, he says, "for both Poles and Italians, the disadvantage is almost 40 percent of the disadvantage encountered by Black college graduates" (Greeley, 1978: 90). The perceived ethnic opposition to affirmative action goals appears to be at least partially related to the recency in which the screens of opportunity have opened to individual ethnic groups.

The black/ethnic issue was recently highlighted by Foner, e.g., citing personal experiences from his classes in black and American history at City College, New York. The affirmative action discussion almost always divided along racial lines. It seems that one of the obstacles in these discussion had to do with the way the disadvantaged are defined for affirmative action purposes. But more importantly, Foner states, "Today, Italian-Americans, Slavic Americans, and other ethnic groups are aware that they have not fully shared in the advantages of this society. They perceive affirmative-action programs as one more obstacle in the way of their own painful climb up the social ladder in America" (1978: 48). He further indicates that ethnics perceive affirmative action programs as an alliance between upwardly mobile blacks and white business groups. Thus he introduces the notion of a class bias in affirmative action programs, as well as a color bias. The previous points are helpful in understanding ethnic opposition to affirmative action, but are not without their own weakness. On the question of color, Hispanic Americans, who are officially classified as white are among the target population for affirmative action programs. Thus, class as well as race are at least implicit in affirmative action programs. However, the class issue is much more complex and may seem to be central at present because most affirmative action conflict to date has centered around university admissions policies. As the conflict spreads to jobs based outside the university the class issue may diminish in importance. Nevertheless, the argument that class rather than race accounts for racial income and employment differential can be used against affirmative action programs.

ECONOMIC INTEGRATION AND THE DECLINING SIGNIFICANCE OF RACE

The decade of the sixties represented a turning point in the life chances of black Americans. It was during this brief interval that many previously existing obstacles were lowered, leading to greater access to both economic resources and social opportunities. Access was not uniformly distributed across all segments of the black population. It has been noted that the greatest advantage accorded males occurred among those who were twenty-five - thirty-four years of age in 1970 (Freeman, 1973: 79). Access accorded black females was more substantial than that accorded males. The income gap between black and white women virtually disappeared during this decade although most of the observed progress occurred after 1966 (Farley, 1977: 362). Needless to say, there are those who view the recent economic progress of black women as less substantial than differences in median income signify. King points out that a disproportionate share of the labor market gains of black women have been confined to the public sector. The private sector, however, and its associated growth industries which have provided opportunities for an increasing number of female workers has failed to provide access to black women (King, 1978: 332-333). Thus both age and sex differences have impinged upon the ease with which segments of the black population acquired access to increased economic resources as a result of the lowering of barriers which had previously confined blacks to a narrow band of the occupational spectrum.

IS SOCIAL CLASS NOW MORE IMPORTANT THAN RACE IN INFLUENCING THE LIFE CHANCES OF BLACKS?

The economic integration of segments of the black population has led to an alteration in the social class composition of the group. The emergence of a substantial black middle class as a result of the lowering of barriers in the economy had led some analysts to conclude that race is declining in significance and that black gains or success in negotiating today's economy is more a matter of class than a matter of race. Wattenberg and Scammon have put the black middle class at 52% of the black population (1973: 36), but Wilson has estimated that the middle class is closer to one-third than one-half of the population. The latter writer is a strong exponent of the thesis of the declining significance of race. The paradigm which he employs to assess the role of race in the economic oppression of blacks leads him to conclude that in a modern industrial economy, class status rather than race is the principal determinant of economic success (Wilson, 1978: 144-154). Wilson specifically states, "The recent mobility patterns of blacks lend strong support to the view that economic class is clearly more important than race in predetermining job placement and occupational mobility" (p. 152). Wilson seems to have arrived at this position by concluding that the state is no longer an oppressor, and that the corporate body is willing to absorb in the corporate economy all blacks who possess the education or skills required to perform efficiently in a modern industrial context. The position of Wilson has been hailed by a growing number of white social scientists, whereas it is being challenged by black social scientists. One reviewer of Wilson's work pointed out a number of flaws in his evaluation of the problem, but acknowledged its strength as the recognition that class is more important than race (Williams, 1978: 147-150). Willie (1978: 15), on the other hand, contends that Wilson's hypothesis is invalid and proposes as a counterhypothesis, "that the significance of race is increasing and that it is increasing especially for middle-class blacks, who because of school desegregation and affirmative action and other integration programs, are coming into direct contact with whites for the first time for extended interaction." The controversy is likely to continue as the Wilson position weakens attempts to secure support for public policies that would further assist blacks in the struggle for economic opportunity.

SOME ADDITIONAL VIEWS ON THE CHANGING SIGNIFICANCE OF RACE

Attempts to specify the individual impacts of race and class on black economic opportunity and the general white response to blacks continue. The results are mixed. One recent study concluded that race is more important than class in influencing the response of white Americans toward black Americans. The authors, Riedesel and Blocker (1978: 564-565) were attempting to partition the role of race and class on social distance. They found through the presentation of a number of vignettes, employing a variety of combinations of social class and race attributes, that "both race and social status explain significant proportions of the variance, but race is clearly more important as a predictor." Additional work along similar lines has been conducted by Hechter (1978: 311). The work of the latter writer more specially concerns itself with the role of ethnic influence on occupational stratification. Hechter concludes that ethnic identity continues to be strong and that such identities influence the division of labor. He states "if ethnicity had little effect on an individual's life chances, its force would surely diminish and there would be one fewer obstacle to the formation of self-conscious classes." Both of these studies attempt, by very different methodologies, to demonstrate the role of race and ethnicity on different facets of racial status. The general conclusion is that race continues to represent a major contributor to social status, interpersonal relationships and occupational differentiation. Neither of these studies directly refutes the Wilson position on the role of race and class on economic opportunity because neither directly confronts the issue that Wilson presented. They do demonstrate the complexity of the situation and indirectly provide evidence which weakens the Wilson position. Nevertheless, one can expect the Wilson thesis to appeal to many individuals with responsibility for administering affirmative action policies.

To be sure blacks are experiencing greater integration into the national economy, but is this integration a function of the declining significance of race? The scale generalization employed by Wilson in support of his position obscures the host of complex interactions that influences black access to individual labor markets. To assess expected corporate behavior in terms of the modal state of economic development masks the role of those local economies whose behavior is incongruent with expected postindustrial employment behavior. If it is assumed that most firms possessing postindustrial attributes provide primary labor market jobs, then one might rationally conclude that race is diminishing in importance as an attribute which influences labor market entry. But even among postindustrial economic organizations, black workers are often utilized to perform tasks that reap advantages to the firm based on racial or ethnic identity. Until recently, for example, there were few black workers holding sales positions within the corporate structure of major firms. The barriers to this occupational category have been slow to erode. Although there is now growing evidence that these barriers are beginning to fall, a closer look at this situation often reveals that black sales personnel are hired to aid firms in capturing a larger share of the black retail market. It appear that in this instance race is very much a factor in the hiring decision and subsequent task assignment. Aggregate assessments based on occupational titles and level of remuneration would lead one to conclude that such penetrations were made essentially on the basis of altered status attributes, an assumption that appears tenuous at best.

The rise of blacks up the corporate ladder is an even more complex proposition. The intimacy involved in high-level decision-making and the social environments in which these decisions are sometimes made tend to result in the exclusion of blacks from these circles. This kind of exclusion grows out of the cultural styles which are manifested at this level of the corporate structure and blacks' general lack of experience in such settings. Thus the cultural style differences which grow out of racial isolation, holding social class constant, lead blacks to be more often excluded from upper-level positions in postindustrial firms. The continuing territorial and social separation of the races will make it difficult for blacks to be considered for positions in the upper reaches of the corporate structure. Hechter (1978: 306) has noted that the high level of ethnic solidarity among blacks, Jews, Asian-Americans, and Italian-Americans is not simply related to their position in the occupational structure, but is partially explained by their high levels of territorial concentration. Thus ethnic identities and cultural styles, which evolve out of a context of territorial or environmental preference or exclusion, weaken the possibility of minimizing the impact of race on occupational penetration.

The conclusion that race has become less important than class in influencing the life chances of blacks appears premature and fails to acknowledge how decisions are made in the hiring and promotional processes. Likewise, the complex ways in which status and ethnicity interact minimize the value of global assessments which attempt to partition the relative strength of each variable on economic progress. And finally, an attempt to confine one's assessment to the economic order independently of the social order appears to disregard the strong linkages between the two.

SUMMARY

Changing economic fortunes growing out of structural changes in the economy have slowed the economic gains of the upwardly mobile segment of the black population. For that group of blacks who traditionally chose to enter nonmetropolitan reather than metropolitan migration streams as a means of enhancing economic opportunity, the rewards associated with this strategy appear to be less certain at present. After more than two generations, blacks have become primarily an urban population. During that same period, the larger urban concentrations themselves have undergone major transformations as a result of increased investments in the development of the residential and workplace infrastructure in the suburban ring.

The transition from an industrial to a postindustrial economy has led to the production of fallout in the form of surplus labor. Older and larger urban locations appear to be losing their power to provide economic opportunity to blacks possessing only limited skills. New labor force entrants who do not possess the attributes required of a postindustrial economy are finding that opportunity is limited in both industrial and rural economies. Declining opportunity in lagging sectors of the economy leads some potential labor force entrants to be attracted to the irregular economy. The vagaries associated with participation in the latter economy mean that there can be no secure economic future for the segment of the population unable to secure employment in alternative sectors. The growing gap in economic well-being among segments of the black population, however, makes it more difficult to fashion policy solutions that are sufficiently comprehensive to overcome the diversity of problems that this population encounters in the job market.

The continued gains on the part of upwardly mobile blacks resulting from favorable public policy are likely to be slowed by opposition from other ethnic groups who view some of the proposed and existing policies as a threat to their economic security and that of their progeny. The limiting of opportunity as a result of slowed economic growth intensifies ethnic and racial conflict and promotes lessened support for civil rights goals.

For those blacks who are trapped in lagging sectors of the economy and for whom absorption into the postindustrial economy is highly unlikely, the future is indeed bleak. The high unemployment levels of blacks in general and black teenagers in particular is indicative of the need to develop policies that will discourage a loss of confidence in the ability of the labor markets to utilize their labor. It has been demonstrated that a policy which promotes the creation of additional government-supported jobs is unlikely to find much support during a period of inflation. In the absence of government-created jobs the private sector will be required to exhibit a good deal of ingenuity in preparing to absorb much of the existing surplus labor. The inability to absorb new entrants to the labor force will possibly lead to an expansion of activity in the illicit sector of the irregular economy. This means a continuation of high crime rates and other deviant activity which is likely to lead to the passage of punitive legislation. Tax payers caught up in the ground swell to roll back taxes through the curtailment of services are most prone to support a reduction in welfare services. The hardening attitudes among that segment of the population which has been most severely penalized by the burden of inflation lessen their sensitivity to the plight of blacks and other low-status minorities whose fortunes were beginning to show signs of positive change prior to recent economic reversals.

Can migration continue to assist in providing economic opportunity through the adaptation of new migrant patterns which reflect sensitivity to regional shifts in economic fortunes? The evidence that Southern metropolitan areas are growing targets for black migrants may indicate that spatial mobility can provide some relief. But it is far from certain that even a significant return flow of limited-skill black migrants can produce the relief sought. As traditional urban settlement patterns are altered and new settlement phases are ushered in that seldom encompass segments of the black population, the ability of the large city to provide opportunity for a growing black population appears to diminish. The rate at which opportunity diminishes is inextricably tied to the actions associated with economic retrenchment.

REFERENCES

BAHL, R., B. JUMP, Jr., and L. SCHROEDER (1978) "The outlook for city fiscal performance in declining regions," pp. 1-50 in R. Bahl (ed.) The Fiscal Outlook for Cities. Syracuse, NY: Syracuse University Press.

BRIMMER, A. (1974) "Widening horizons: prospects for Black employment." Rev. of Black Pol. Economy 4 (summer): 91-115.

――― (1972) "Economic situation of Blacks in the United States." Rev. of Black Pol. Economy 2 (summer): 34-54.

Chronicle of Higher Education (1978) "The opinion of Justice Marshall," July 10.

FARLEY, R. (1977) "Trends in racial inequalities: have the gains of the 1960s disappeared in the 1970's?" Amer. Soc. Rev. 42 (April): 189-208.

――― and A. F. TAEUBER (1974) "Racial segregation in the public schools." Amer. J. of Sociology 79 (January): 888-905.

FONER, E. (1978) "Bakke reconstructed" Chronicle of Higher Education (March 13): 40.

FONER, P. S. (1974) Organized Labor and the Black Worker in 1619-1973. New York: International Publishers.

FREEMAN, R. B. (1978) "Black economic progress since 1964." Public Interest 52 (summer): 61-64.

――― (1973) "Changes in the labor market for Black Americans, 1948-1972." Brookings Papers on Economic Activity 1: 67-131.

FRIEDLANDER, S. L. (1972) Unemployment in the Urban Core. New York: Frederick A. Praeger.

FUSFELD, D. R. (1973) The Basic Economics of the Urban Racial Crisis. New York: Holt, Rinehart and Winston.

GEORGAKAS, D. and M. SURKIN (1975) Detroit: I Do Mind Dying. New York: St. Martin's.

GREELEY, A. M. (1978) "Ethnic minorities in the United States: demographic perspectives." Int. J. of Group Tensions 8, 2: 64-98.

HECHTER, M. (1978) "Group formation and the cultural division of labor." Amer. J. of Sociology 84, 2: 293-319.

HENRY, J. S. (1978) "Lazy, young, female, and Black: the new Conservative theories of unemployment." Working Papers (May/June): 40-58.

HOGAN, D. P. and D. L. FEATHERMAN (1977) "Racial stratification and socio-economic change in the American North and South." Amer. J. of Sociology 83 (July): 101-126.

JONES, M. H. (1978) "Black political empowerment in Atlanta: myth and reality." Annals 439 (September): 96-117.

KING, A. G. (1978) "Labor market racial discrimination against black women." Rev. of Black Pol. Economy 8, 4: 325-335.

LEGGETT, J. C. (1968) Class, Race, and Labor. New York: Oxford University Press.

National Urban League Research Department (1976) Q. Econ. Report of the Black Worker 7 (August).

PRESTON, M. B. (1977) "Blacks and public policy." Policy Studies J. 6 (winter): 245-255.

RIEDESEL, P. L. and J. T. BLOCKER (1978) "Race prejudice, status prejudice, and socio-economic status." Sociology and Social Research 62 (July): 558-571.

SACKS, S. (1977) "Changes in manufacturing and retail employment in medium sized cities" pp. 113-151 in H. J. Bryce (ed.) Small Cities in Transition. Cambridge, MA: Ballinger.

SILVER, C. B. (1973) Black Teachers in Urban Schools. New York: Frederick A. Praeger.

SMITH, J. P. and F. WELCH (1978) Race Differences in Earnings: A Survey and New Evidence. Santa Monica, CA: Rand Corporation.

STERNLIEB, G. and J. W. HUGHES (1977) "New regional and metropolitian realities of America." J. of the Amer. Institute of Planners 43 (July): 227-241.

THOMPSON, W. (1977) "The urban development process" pp. 95-112 in H. J. Bryce (ed.) Small Cities in Transition. Cambridge, MA: Ballinger Publishers.

TURNER, J. H. and R. SINGLETON Jr. (1978) "A theory of ethnic oppression: towards a reintegration of cultural and structural concepts in ethnic relations theory." Social Forces 54 (June): 1001-1018.

VILLEMEZ, W. J. and C. H. WISWELL (1978) "The impact of diminishing discrimination on the internal size distribution of Black Income: 1954-74." Social Forces 56 (June): 1019-1034.

WATTENBERG, B. J. and R. M. SCAMMON (1973) "Black progress and liberal rhetoric." Commentary (April): 35-44.

WILLIAMS, W. E. (1978) "Race and economics." Public Interest (fall): 147-154.

WILLIE, C. V. (1978) "The inclining significance of race: commentary." Society (July/August): 10-15.

WILSON, W. J. (1978) The Declining Significance of Race. Chicago: University of Chicago Press.

11

Anglo Retrenchment and Hispanic Renaissance: A View from the Southwest

WARNER BLOOMBERG, JR.

□ "RETRENCHMENT"–CUTTING DOWN, CURTAILING, reducing–remained the dominant theme for domestic political rhetoric as the post-election fall faded into an early winter in many parts of the United States and 1978 drew to a fitful close. The President warned the mayors that the forthcoming federal budget would be the very embodiment of retrenchment, with most urban-oriented programs at best constrained to current levels, some cut, and little money for new starts. He thereby echoed demands for diminishing the cost of the public sector first implemented at the local level by a spate of tax-cutting and bond-defeating campaigns in a variety of municipalities, school districts, and states, of which California's Proposition 13 was the most publicized. Federal retrenchment will now reverberate back through the state and local systems.

During such times it is difficult to disengage from the immediacy of current statistics on inflation, employment, income, and the unmet needs of those in the lower reaches of the economy for whom a winter of increased hardship is in store. What follows is an effort to set the concern for urban institutions within a perspective which looks backward for several decades to see where we have been going and also attempts a necessarily speculative anticipation of the future. It is an attempt to discern the pattern of a system with its feedbacks and circular relationships rather than to establish any particular sequences of causal relationships.[1] It will be argued that the "retrenchment" is quite selective, that it impacts most upon those institutional systems that benefit the minorities and the poor within the central cities. Public retrenchment protects the private sector institutions most valued by the more affluent and predominantly white (Anglo) population who resides in the increasingly dominant outer rings of the still expanding metropolitan areas. Indeed, the private sector institutions of the relatively decentralized outer metropolis are becoming, if anything, more firmly entrenched as a result of the constraints and cutbacks now faced by urban public enterprises and delivery systems.

It seems likely that this may be the prevailing pattern well into the 1980s. It is quite possible, although far from certain, that toward the end of that decade Hispanic-Americans, in coalition with other minorities encapsulated within the central cities, will be inclined and able to put in place new policies for revitalizing their muncipalities and communities. But such a "renaissance" would not restore the central city of the past, nor would it embody the tastes and ambitions of the elites who for the most part will continue to reside in the outer metropolis.

If such a change were to occur, it would most likely appear earliest and be carried farthest in automobile cities of the Southwest, where by 1990 Hispanics/Latinos will constitute with largest bloc of the resident population. Ironically, many of these cities are at the center of newer metropolitan areas in which the shift of affluent Anglos and political and economic dominance to the outer metropolis has been unfolding much later than in the older urban-industrial centers of the East and Midwest. The continuing influx of Latinos, primarily Mexicans and Mexican-Americans, is undoubtedly hastening that transition. The continued demographic and economic growth of the Sunbelt region would seem to assure that patterns which emerge there will have substantial impact on national culture and policies, although not as much as some regional megalomaniacs seem to expect. A view from the Southwest, with its parochialism properly restrained, may therefore be predictive for the nation. (The one parochialism which is indulged without restraint here is the use of "Anglo" to mean whites in general, a usage which is spreading rapidly in the region's mass media.)

FROM PAST TO PRESENT:
THE EMERGENCE OF THE "DUAL METROPOLIS"

The image of the dominant central city has long prevailed in both popular thought and professional literature. By World War I it was clear that the influence of this city was transforming hinterland towns into satellite cities (Taylor, 1915). Suburbs, which began developing at the turn of the century, had become a proliferating part of the urban hinterland after World War I (Douglas, 1925). The depression cut off much of this growth, and the rather simplistic distinctions between central city, suburbs, and satellites seemed to hold up well through the development hiatus of World War II (Harris, 1943). The resurgence of peripheral expansion which followed the war promoted the anticipation of "conurbations" which eventually would merge at their peripheries, in the end weakening but not eliminating the central business district (Davis, 1955). But the established conceptualization of a metropolitan area with a dominant central core, almost always the central city, and one or two outlying rings of suburbs and satellites, continued to be the framework used by most analysts for aggregating their data (Dobriner, 1963), partly because it facilitated comparisons along time-lines, partly because no compelling alternative was available.

TOWARD A NEW IMAGE OF NEW REALITIES

Yet it is clear that the increasing pace and selective character of peripheral expansion have produced an urban system which is ever more poorly represented by the old model. Berry and Kasarda (1977: 174) assert that "the emergence of a new urban form is unmistakably upon us ... a radically changed, unbounded metropolitanization that makes urban-non-urban distinctions meaningless." As part of this new form,

> a multinode, multiconnective system has replaced the core dominated metropolis as the basic urban unit. Many suburban and exurban areas now provide all essential services and numerous specialized services formerly concentrated in the city core; new outlying locations provide for shopping needs, jobs, entertainment, medical care, and the like. It is these and other facets of ecologic change such as the spontaneous growth of the exurban periphery that today combine to constitute the daily urban

systems that far transcend the Bureau of Budget's more narrowly defined SMSAs. [Berry and Kasarda, 1977: 304]

The areas stretching outward from the old central city, increasingly filled in with new developments, shopping centers, and business and industrial parks, are also predominantly white collar, Anglo, and above average in education (Schnore, 1972; Smith, 1970). Both blue-collar and white-collar employers have also migrated outward, the latter to a greater degree than the former (Colenutt, 1972); but residences for those of working-class income have remained relatively scarce, and blacks have remained grossly underrepresented in the outer metropolis regardless of occupation (Kain, 1968, Berry and Kasarda, 1977).

The view inward toward the core of the central city provides striking contrasts. The poor and the nonwhites constitute an increasing proportion of the central cities' population (Downs, 1968). Inner-city unemployment and underemployment remain high, refractory to policies which operate to increase employment in industries that are no longer present where poverty abounds and often inaccessible to the unemployed poor. Many residence of poverty areas become recurring, if not permanent, recipients of public sector assistance as "the economics of uselessness are replacing the economics of exploitation" (Turner, 1970: 3).

As a result of these trends, many central cities can realistically be conceptualized as "functionally separated from the rest of the SMSA . . . increasingly assuming the characteristics of depressed areas" and experiencing "a decreasing amount of spatial interaction with the suburbs" (Colenutt, 1972: 97). Although many people of wealth and influence still reside in the central cities, and far more of moderate but comfortable means, and although a majority in almost all of the larger cities are still Anglos, the old city as a social entity has changed in irreversible ways in the United States. It is no longer the rich center that once so clearly dominated the bedroom suburbs and modest satellite cities scattered outward from its boundaries in an unintegrated patchwork with the rural hinterland. Now inner means thinner in all but population density. The total metropolitan region may usefully be thought of as divided between two increasingly distinctive metropolitan areas, an "outer metropolis" whose character is set by its relatively affluent, Anglo majority (and will therefore also be referred to as the "Anglo metropolis") and an "inner metropolis" which continues to suffer corrosion and degradation not only as a physical habitat and community, but also politically and economically. Such an image is, of course, a gross oversimplification of the complex structuring of contemporary urban America. But that shortcoming will not diminish its utility for directing attention to some of the underlying dynamics which have produced that assault on the urban public sector for which California's Proposition 13 seems so paradigmatic.

THE OUTER METROPOLIS AS A SYSTEM

One reason that we have not thought of the outer part of the dual metropolis as one system probably is that its coherence and integration are not expressed or achieved through a governmental structure, such as the municipality or county through which the inner metropolis is governed. Suburban areas have been characterized as a "Balkanized" hodgepodge of municipalities, school districts, and other more specialized units. But true "Balkanization" involves diverse ethnic and class groups distributed territorially and organized politically so that each can seek to preserve internal homogeneity and to deal with neighboring territories as natural adversaries. Rarely is this the case as one moves from suburb to suburb. There are some working-class suburbs, usually just beyond the central city's boundaries, and a few Black suburban settlements (Rose, 1978). But for the most part there is almost no sense of anything changing except that some are more affluent than others.

Multilateral coordination derives easily from this sameness, and from the provision of some services by overlying governments and agencies (Walton, 1968; Winter, 1969), such as counties and special districts for water and sewage and sometimes even private organizations that cross jurisdictional lines (Liebert, 1976). Critics have long argued that such a "nonsystem" should founder, lacking as it does the integration of an overall administration and decision-making structure. Instead, the overlapping pattern of jurisdictions, shopping areas, and associational ties has served as a network capable of knitting together enough for their purposes the majority of those inhabiting the outer metropolis, with their limited socioeconomic and cultural diversity and their broad normative consensus about the value of a local government of "limited liability" as well as limited territory (Greer, 1962a, 1962b).

To say this is not to put forward an exaggerated view of the solidarity and cohesiveness of the Anglo metropolis. In addition to some nonwhite and working-class suburbs, older satellite cities are sometimes minor replicas of the central city with its ethnic and racial minorities and its pockets of poverty. But such exceptions to the general milieu of the outer metropolis can act only within their own jurisdictional boundaries on behalf of policies different from those serving the interests of the more affluent Anglos. The very absence of one overarching government with a broad range of functions serves to contain the influence of "deviant" policies.

The pluralism which is valued in the outer metropolis uses the private far more than the public sector as the means for its expression. It involves variation, reflecting not so much the old differences in urban populations of ethnicity and socioeconomic status, as the diversity of organizing motifs now made available through leisure, consumership, and selective sociability as vehicles for elaborating images and intentions for self and household. Probably only a numerical minority have rounds of life focused fairly tightly on a singular fascination—golfing, gardening, boating, politicking, swinging, some sectarian commitment such as a church or a holistic self-development group, and so on; but there is a good deal of tolerance and even admiration for such intentional subcultures as long as their devotees do not become abrasive advocates or intrusive missionaries. Those who are not caught up in one or another of these subcultures share in the more generalized, leisure-oriented lifestyle of Anglo affluence to the extent that their incomes permit. The key to achieving lifestyle goals and interests is thus disposable income to purchase goods and services appropriate to the pursuit of the valued activities, environment, and associations. Income allocated to government is understandably perceived as diminishing that pursuit.[2]

These are "intentional" lifestyles. They do not derive from ethnic traditions received during primary socialization. Neither do they reflect such external circumstances as poverty, living on a farm instead of in an urban setting, or having to accommodate to patterns of togetherness and apartness and residential mobility that go with some occupations, such as the military. They are both fully voluntaristic and "now"-oriented, like the suburban homesite and community where one dwells only so long as each facilitates the chosen round of life, or that round remains one's choice. Commitments to such lifestyles may thus be intense without being persistent. Finally, even the more extreme forms of intentional lifestyles, such as "counterculture" communities and the variety of exotic non-Christian religious cults which enrage middle America, seldom appear in the Anglo metropolis. Thus, the pluralism of this part of the dual metropolis requires few public expenditures either to facilitate its expression or to resolve conflicts which it might generate.

KEEPING THE ANGLO METROPOLIS "PURIFIED"

Sennett (1970) looking at the emergence of this sociocultural milieu from the perspective of philosophical anarchism has suggested insights with particular relevance to this analysis. What we are seeing, Sennett argues, is an effort to preclude those sources of dissonance and unexpected conflicts which make the future of one's community and own life unpredictable, to exclude changes which happen to one rather than being chosen. This effort to "purify" one's social milieu leads to efforts to eradicate, not all differences, but those which can produce confrontations or other unknown social situations full of surprises and challenges. Few, if any, such threats to life's predictability inher in the intentional lifestyle pluralism of the outer metropolis. One need not fear race riots; politics are never imbued with the obsessive sentimentalities of ethnic traditions; nearby slums do not haunt the civic consciousness or disturb its sense of security with their restiveness, whether actual or imagined. Order and predictability are enhanced by the stable character of the environment: residential developments, shopping centers, an adequate allocation of church sites. And in the fact of that high degree of privatization of the round of life which affluence facilitates, there is a myth of community which rests upon the underlying similarity of the citizenry rather than upon an involvement in the continuing resolution of problems which is compelled by real diversity.

That environment would be likely to satisfy the individual who can afford to purchase or rent homes within the price range of the outer metropolis, at least as Berry and Kasarda (1977: 81) characterize him:

> In the search for self-identity and security in a mass society, he seeks to minimize disorder by living in a neighborhood in which life is comprehensible and social relations predictable. Indeed, he moves out of "his" neighborhood when he can no longer predict the consequences of a particular pattern of behavior or patterns of behavior of his neighbors. He seeks an enclave of relative homogeneity: a territory free from status competition because his neighbors lead similar life styles . . . a safe area, free from status-challenging ethnic or racial minorities.

Maintenance of such a "purified" social order requires sustaining limitations on diversity and the simplification of issues. This means that the central city must continue to "contain" the conflicts and complex problems arising from the clash of class interests, disparate ethnic-racial subcultures, and deviant subcommunities. In this sense the inner metropolis must serve as a "holding area" for the poor, the nonwhite minority enclaves, and less numerous but still disturbing groups such as "gays" and unusual religious cults. The main mechanism for achieving this is the reliance on disposable personal income for support of households and lifestyles where rents are high and real estate continues to escalate, public transit often is almost nonexistent, mercantile outlets for cheap goods (however shoddy) are few and far between, and public social service agencies and facilities are even rarer. Remnants of racial discrimination in the housing market add to the selective barriers, so does the absence of political traditions and mechanisms which give unpopular ideological and lifestyle minorities in the cities something of a "fighting chance."

Thus, the relatively small proportion of minority group members who has the financial means for full participation in the predominantly private sector urban institutions of the Anglo metropolis is in fact underrepresented among its residents. And it is likely that most of those non-Anglos motivated and able to overcome any resistance by Anglo sellers, real estate agencies, and lenders, and willing to be relatively isolated from their ethnic communities, will already display lifestyles compatible with those of their white neighbors. Indeed, dress, home decor, and even political postures symbolizing a Black or Latino identity can be viewed as "charming" and easily incorporated into the intentional lifestyle pluralism just described. It is therefore not surprising that the proportion of Anglos who assert that they would stay put if Blacks moved in next door has increased from 55% in 1963 to 84% in 1978, while those who say they would move if Blacks entered their neighborhoods in great numbers has declined from 78% to 45% during the same fifteen years (Gallup, 1978a). The reality is a United States at least as segregated residentially by race now as it was two decades ago, and perhaps more so (Rose, 1977). Most of the Anglos answering the pollsters' question about a large influx of Blacks know that they simply will not face such a situation. Their responses can be pure good intentions. Only those Anglos on the peripheries of expanding urban ghettoes have experienced such a change, and they have almost universally fled. But they constitute only a very small proportion of the white majority in this highly segregated society. The vast majority in the purified community know the question is purely hypothetical.

THE INNER METROPOLIS

If the outer metropolis coheres better as a system than critics of its polynucleated, decentralized format believe, most contemporary central cities are patently far less integrated, as well as far less powerful, than the conventional wisdom purports. The image of centralized power persists in the skyline and amenities of the central business district, which has been enjoying something of a building boom of new offices and luxury high-rise residences in many cities in recent years. Here are the offices of major corporate business enterprises and highly successful professionals, plus mercantile establishments, cultural and ceremonial centers, and recreational facilities utilized by the more affluent, some of whom also reside in apartments and condominiums within this area, both of older vintage and of more recent, "urban renewal" origins. Yet most of this, while obviously at the heart of the central city's territory, is "of it" only in very limited ways because legalistic jurisdiction for purposes of zoning, taxing, and property protection can be a very minimal sort of systemic integration. Much of the prestigious physical plant is owned by supralocal corporate organizations with only those commitments to the locality required by practical politics and public relations. A majority of those who use the area— employees, customers, and clientele—reside in the Anglo metropolis and have no regard for the central city apart from the maintenance of what serves their purposes. And it is doubtful if their contributions to the central city's economy through trade and taxes balance the costs to the public sector of servicing those transactions and exchanges and the facilities they require (Kasarda, 1972).

Usually stretching outward from this elitist center, sometimes penetrating into its most dilapidated side streets, is a vastly different physical environment and social milieu: impoverished whites, elderly, unskilled new arrivals, failures in the competitive job market, and racial minority communities whose more affluent members stretch the territory of ghetto and barrio toward the receding perimeter of the Anglo world and its more desirable housing stock. Also within the inner half of the central city are persisting areas inhabited by the lower-paid blue-collar and white-collar workers. Few of those living in these parts of the central city make any use of or receive any benefit from the highly developed business and cultural facilities of the central districts and similar subcenters elsewhere within the city. But their declining neighborhoods have been vulnerable to disorganization and even destruction to facilitate the expansion or servicing of business and affluent residential developments. And their needs for property protection and other services have usually been given lower priority than meeting such needs in the areas of higher income and status.

As the proportion of the central city population who is either of low status, poor, or both, increases, those of higher status and income who do not move to the outer metropolis become concentrated in the peripheral areas and, in smaller numbers, the islands of affluence within the inner core. Analysts considering the urban violence of the sixties anticipated the patterns increasingly evident by the end of the seventies:

> It is logical to expect the establishment of the "defensive city," the modern counterpart of the fortified medieval city, consisting of an economically declining central business district in the inner city protected by people shopping or working in buildings during daylight hours and "sealed off" by police during nighttime hours. Highrise apartment buildings and residential "compounds" will be fortified "cells" for upper-middle and high-income populations living at prime locations in the inner city (Mulvihill et al., 1969: XXV).

Available evidence, both systematic and impressionistic, indicates that a majority of the residents in such areas and locations share many more lifestyle preferences, values, and political judgments with their economic and status peers in the Anglo metropolis than they do with the poor, the minorities, and the blue-collar and lower-income white-collar denizens of the rest of the central city. And the resources, organization, and political skill which can be utilized by this upper-middle and upper-strata citizenry still enable them to exert substantial influence on the policies and practices of the local government.

For the central municipality is itself poorly integrated as a decisional system (Long, 1958; Greer, 1962b). It is an "arena" in which contending interests deal and are dealt with through the power brokerage of municipal executives and the professionalized authority of governmental bureaucracies (Kotter and Lawrence, 1974). In short, it is a mechanism for resolving, usually without violence, the conflicts arising out of the Balkanized population of the central city. Just as the business and cultural centers are functionally adjuncts of the Anglo metropolis, many of the more affluent Anglo residents of the central city frequently represent a common policy orientation with their counterparts in the outer metropolis. This, too, contributes to the new balance of power between the inner and outer segments of the dual metropolis. For the role played by the majority of the affluent Anglos in the arena of Balkanized central-city politics reduces the chances that minorities and excluded whites will be able to make the municipal government a vehicle for discovering and developing their common interests.

THE "DEPENDENT ECONOMY" OF THE INNER METROPOLIS

The use of the public sector to deal with the economic and environmental problems faced by minorities and low-income residents of the inner metropolis is surely one of the most compelling of these common interests. For the poor the public sector is the major source of such limited assistance as is available: cash transfers, food stamps, public clinics, public housing. For many hope of employment in the private sector appears to depend frequently upon publicly funded training and employment underwriting, while the public sector itself, especially at the local level, has been an important source of urban jobs for racial minorities. Affirmative action programs also have relevance for the economic opportunities of minorities, but this brings them into a clash of interests with their Anglo counterparts.

But a large number of the employed urban poor and near-poor work in occupations in what has been called the "periphery sector" of the economy (Averitt, 1968; Bluestone et al., 1973; Beck et al., Tolbert 1978). Periphery sector firms are characterized by small size, seasonal and other variations in product supply and demand, labor intensity, weak unionization, and low assets. Such firms have increased as a proportion of the inner-city economy as the industries, larger businesses, and professional establishments of the dominant "core sector" have moved outward into the Anglo metropolis. Employees within the periphery sector are likely to have substantially lower incomes than those in the core even when age, education, and experience in the workforce are held constant (Beck et al., 1978). The disproportionate allocation of nonwhites into this sector by virtue of location as well as employer discrimination helps to explain the continuing income disadvantage of employed minority workers relative to employed Anglos.

The business and manufacturing areas of old inner cities, deserted by so many of the large corporate enterprises of the core sector, are characterized not only by many marginal small manufacturers paying low wages to nonunionized workers but also by a multitude of enterprises in mercantile and service lines attempted by residents of the minority enclaves. Unlike the corporate oligopilies, they operate in an open, competitive environment, but their efforts are often inadequately capitalized, their technical knowledge and relevant "connections" limited. Most Black businesses are found in such areas, along with the characteristics abundance of unoccupied rental spaces, and they experience two to three times as many failures as Anglo entrepreneurs (Brimmer, 1969). Most aid to minority business efforts has come from the public sector either directly or through the subsidy of firms located in the inner city and willing to engage in such aid. Governmental agencies also have financed development corporations as another means of invigorating the low-income areas' share of the economy of the inner metropolis. The dream of an increasingly independent "Black capitalism" has thus proved illusory, as have the job training programs that were supposed to reduce minority unemployment substantially by moving the unemployed into jobs with a future in the core sector of the economy of the inner metropolis. For many Blacks, and members of other minorities as well, expansion of governmental agencies and programs has offered more accessible and rewarding job opportunities than either minority enterprises or the increasingly distant factories and businesses of the Anglo-dominated core sector.

For the city's low-income residents generally, supportive public sector interventions have been the overwhelming source of such improvements in the quality of their lives as mass transit, improved housing, revitalized residential environments, neighborhood clinics and recreational centers, and the like. Obviously, need continues to outrun effort by a large measure. Public sector institutions also are the most likely, and in many cities probably the only, substantial source of services oriented to the cultural traditions and particular needs of minorities, apart from that limited supply generated by each group's own entrepreneurs and professionals. Schools with bilingual programs, community orientation programs for the police, medical and social service clinics and outposts within the neighborhoods and staffed from within the community can be responsive to the values and traditions of minority groups (March, 1968; Marr, 1972).

The lower-income and minority groups of the inner metropolis experience a dependency on the public sector in direct and obvious ways which the more affluent do not. The latter, like their counterparts in the outer metropolis, are in fact beneficiaries of a wide variety of governmental contributions to their economic well-being, ranging from VA and FHA subsidies of home mortgages to all the state and federal supports for firms in the core sector of the economy. But these are institutionalized and thus enjoy the invisibility of being part of what is "normal." Aid to the poor and supportive direct interventions for minority communities are viewed as confirmations of their dependence and questionable treatments subject to special political vulnerability.

WHO GETS RETRENCHED?

The preceding analysis of the emergence of the duality in contemporary metropolitan areas sets the stage for the so-called "retrenchment" which is illustrated, perhaps epitomized, by the passage of Proposition 13 in California. For reasons which should now be clear, reducing that part of their income which supports the public sector seems to a great many homeowners who predominate in the Anglo metropolis to protect their lifestyles, which depend so much more on expenditures in the private sector than on direct support and services from government. By cutting down on taxes they are defending their chosen way of life in a time when consumer prices rise steadily. It is a kind of "entrenchment" to protect the urban institutions they value: expensive homes and condominiums with big mortgages, expansive shopping centers, commercialized recreation, a wide variety of private clubs and associations, and so on. If this results in retrenchment of public sector institutions on which the poor in general and minorities in particular depend, that is not only a price which must be paid, but it is one which many seem to feel is quite justified.

As one commentator put it:

> Suburban Californians are a nervous breed of the "newly arrived." They suffer the neurosis of having too much and not knowing how well they have it. Fear of displacement makes the

newly arrived rather unpleasant people to live with.... The nervous suburbanites ask themselves every morning what will save them from labor unions, taxes, blacks, Cesar Chavez, the women's movement, busing, illegal aliens, flouridation, and worst of all, government. They waited for the message that would preserve them from these evils, and Howard Jarvis gave them the message. [Blaustein, 1978: 21]

Thus, relative deprivation combines with the predeliction of the purified community. Affluent consumers, coveting such joys as video cassetts and recorders, hardly experience the crushing blows which inflation brings to the poor, but they feel greatly threatened nonetheless. Some simple statistics are quite suggestive. Real median family income in the United States increased by 34% between 1960 and 1970, but by only 4% from 1970 to March 1978 (Bureau of the Census, 1978). However, between 1961 and 1973 American families increased their spending on food by 8%, on housing by 37%, and on transportation by 50% (Bureau of Labor Statistics, 1964, 1978). Housing and transportation are the most important components of rising costs for the residents of the Anglo metropolis. It is not so much that one's standard of living is declining, but that its anticipated continuing advance may be cut off.

If taxes are seen as a source of that threat, the appropriate reaction seems obvious. Whatever their biases, the citizenry of the Anglo metropolis are relatively well-informed. They have heard that government salaries have been rising faster than private sector personal income, and they know that transfer payments have soared. That both of these patterns may be appropriate to the situations of those affected is simply not on the agenda of outer metropolis concerns. Some proportion of the residents of the outer metropolis hold jobs with governmental agencies or in private firms with government contracts. But the cost of government and the consequent tax burden may strike even those whose incomes are part of the total public sector expenditure as a threat to be reduced, especially if the reduction seems likely to have its major impact elsewhere.

If the California action is to be understood as paradigmatic, it is important to discriminate between what can be emulated elsewhere and what has been relatively unique in the California situation. The latter includes the build-up of a huge state surplus generated by both continuing economic expansion in the state and inflation, which the legislature failed to redistribute because of political wrangling. As property values in California, primarily in the outer metropolitan areas, inflated at a rate which can only be described as excessive to grotesque, whereas tax rates were not adjusted accordingly, suburban homeowners understandably became increasingly fixated on the costs of government, even though theirs, except for often rather expensive school systems, was usually minimal at the local level. To this add the California tradition of extensive use of the referendum, and the situation was finally ripe for a proposal which had been touted for some years by Howard Jarvis, a wealthy retired businessman who now headed the Apartment Association of Los Angeles County. His collaborator in the state legislature was Paul Gann, a retired real estate salesman.

Besides rolling back property tax rates and restricting future levies, the referendum established a two-thirds majority in both houses of the state legislature for passage of any other state taxes. Intended but not accomplished in subsequent legislative reactions was a set limit on total governmental spending, but Governor Brown's requirement of priorities for a 10% budget cut from all state agencies suggests a move in that direction through administrative action. Moreover, local governments expecting future "bail-out" funds from the state surplus to replace some of the losses from property taxes must also undertake the budget exercise. Policies and practices such as these can be adapted by other states and other levels of government.

The differential impact on the inner and Anglo segments of the dual metropolis is evident. So extensive an assault on the property tax is not only regressive, but it benefits proportionately larger numbers where there are more homeowners. Property tax limits will also provide protection for those who reside in the outer metropolis but who own, both individually and in corporate form, the valuable real estate concentrated in the central business districts. Thus, owners of real estate are relieved of the fear of having to help finance social services to residents of the inner metropolis.

Indirect benefits to home renters through rent reductions have been notable by their scarcity, and rent increases have continued to appear. It is too soon to see if the purported stimulus to business investment will become evident and if it will increase employment accordingly. But firms in the core sector will benefit far more than those in the periphery sector because the property of the former is more valuable by both location and capitalization. Since they also tend to be technology- rather than labor-intensive, to require high skills, and to be located outside of the inner city, increased employment opportunities would be unlikely to reach the unemployed and underemployed of the inner metropolis in any case.

How are benefits likely to flow through such a filtering down as may take place. Obviously, members of taxable private associations, such as country clubs, will benefit. Other beneficiaries are suggested by two actions that coincide with the time of this writing. First, Wells Fargo and Co., a banking and investment firm, will distribute 1, 2 million dollars of its tax reduction to charitable organizations (with beneficial tax implications): one-third to public television stations, one-third to United Way (most of whose member agencies serve primarily middle- and upper-middle-income clienteles), and one-third to housing rehabilitation groups, a disbursement which will be of aid to low-income areas (and will keep the poor in the inner metropolis). In sum, the company distribution of the surplus appears weighted toward benefiting upper- and middle-income persons. Second, the owners of two small-business parks in suburban communities south of Los Angeles, near Long Beach, will give rent rebates to its tenants. Neither market competition nor bargaining skills are on the side of inner-city residents attempting to achieve rent reductions. Overall, it is difficult to imagine a tax-cutting approach which would be of greater aid to the mainly private sector institutions of the outer metropolis.

The obviously negative impact on the public sector is equally clear, and of far greater consequence for the central city with its minorities and poor, than for the Anglo metropolis. As governmental budgets are cut, those considered "essential" are most easily protected: police, fire protection, sanitation, and basic legislative and administrative bodies. They also have had the highest priority for "bail-out" funds from the state. Thus, the high priority of property-related services helps insure continued protection of private property, the sine qua non of the dominant class, both within and outside of the central city. The most vulnerable agencies are proving to be those most involved in meeting the needs of minorities and the poor for employment, housing, recreation, education, social services, and a variety of specialized support functions oriented to the life situations and cultural values of particular ethnic groups. Among the easiest targets for budgetary surgery are matching funds for federal programs, many of which have been questioned by local conservatives from their inception and would never have been adopted without the subsidy. Cutbacks in welfare have been advocated with renewed vigor at every level within the state. Opposition to bilingual programs in the schools has been similarly reinvigorated. School budgets have been reduced by diminishing counseling services and a variety of modest innovations, losses far less consequential for the relatively privileged offspring of the outer metropolis than for the children of the inner city. In short, the limited public sector of the Anglo metropolis will suffer less total damage precisely because the quality of life there depends so much less upon its intervention and support than is the case in the inner metropolis, at least in the areas where most minority groups and low-income households reside. There is, then, little doubt about who is being retrenched.

THE FUTURE OF RETRENCHMENT

If the argument is correct, what is being called retrenchment is likely to last a long time, even though some of its evident excesses may be corrected. For it amounts to a major consolidation of the perceived interests of the majority in the now dominant outer metropolis. It will further solidify the existing divisions between a relatively affluent Anglo-dominated area, not all of whose residents are white, and a central city which serves both to provide desired adjuncts to the business and recreational life of the Anglo world and to contain minorities and poor whites within the ghettoes, barrios, and interstitial areas of the inner metropolis. Indeed, what is emerging might well be called a policy, not of retrenchment, but of entrenchment.

The brief and minimal "war" on poverty of the Johnson administration provided a test run of the federal government's ability to serve as a countervailing power to such entrenchment. But the political context for the resurgence of even that much effort to deal with persisting poverty and solidifying racial and economic segregation does not seem in the offing. In spite of the symbolic readiness to move in such a direction expressed in the platform ideology of liberal Democrats, the Carter administration has made evident its understanding that the fight against inflation precludes any such initiative on a meaningful scale. Moreover, most Anglo voters apparently believe that public sector cutbacks will harm only the "chiselers" and the bureaucrats who coddle them. The perceptual and judgmental chasm between whites and nonwhites is well-illustrated by the Gallup poll which disclosed that 73% of the white respondents believed that Blacks had as good a chance as whites to gain any kind of job for which they were qualified, a view held by only 38% of the Blacks (Gallup, 1978b). Again, 71% of the whites polled believed that Blacks were treated the same as whites in their communities, a sanguine outlook shared by only 34% of the Blacks.

Such a disparity is hardly surprising. Few whites in general, and even fewer of those who live in the Anglo metropolis, ever actually view the landscape of urban economic depression where so many minorities live, unless they watch a TV documentary on the subject. And those who encounter nonwhites in the predominantly Anglo workplaces of the core sector are more sensitive to what they see as the inequities of affirmative action programs in hiring and promotion than to the continued underrepresentation of minority group members. Nothing currently visible on the political horizon indicates that liberal Anglos are going to be able to move the Anglo majority in new directions.

What seems likely in the era of entrenchment are recurring episodes of inadequate recycling of such programs as job training, CETA, various versions of "compensatory" education, minimal efforts to produce adequate housing, reforms of welfare that do not restructure income, and the like. This may anguish the liberal conscience, but it will keep workable the kind of society now in place in the dual metropolis. The weakening of the public sector at the local level amounts to diminishing the power base of the politically oriented of the inner more than that of the outer metropolis. Thus, until something happens to alter the beliefs or political balances which are producing retrenchment/ entrenchment, it should continue as this nation's "urban policy."

THE "RENAISSANCE" OF RETRENCHMENT

A counterpoint to this analysis has recently been offered by Allman (1978) in a popularized report on work undertaken through the Third Century America Project.[3] U. S. cities are seen as focal points in an international system of both trade and migration whose trends contribute importantly to a continuing reshaping of the urban environment and metropolitan structure. Allman argues that federal programs of all sorts have in fact substantially increased over the past two decades to the point that large proportions of city general revenue are directly derived from federal aid. Moreover, private investment in central city construction continues to increase, and with it jobs in the intellectual and technically sophisticated occupations which will be housed in the new additions to the skylines of both central and peripheral business districts. With this comes the conversion of some of the decayed areas into refurbished or reconstructed residences for upper-middle and upper-strata households, including many of those minority group members who have made into these valued occupations and statuses. Moreover, Allman argues, most of the urban programs directed to the problems of the poor and the excluded minorities have been ineffective and even counterproductive.

The result of all this is "an odd renaissance," as Allman puts it, in which cities become both richer and poorer. If this analysis is correct, increasing portions of the old inner city will be reclaimed for those who are well off and want to live near to where they work and play in an era of escalating commuting costs. But the population of poor and near-poor also will be increasing due to Third World migration, so that ghettoes and barrios, many of them communicating internally largely in their native tongues for many years to come, will also be expanding. "Urban renewal" will continue to push them from the core area along sectors of present expansion outward. The "slumming" of inner suburbs is already marking this transition in some of the older, larger metropolitan areas. Inner-city affluence will be marked by more private schools (aided by tuition vouchers) in renewal neighborhoods and by the boutiques and restaurants locating where marginal manufacturers and discount stores once employed and sold to lower-income residents.

Like all scenarios for the future, Allman's is highly speculative. It conveniently disregards the evidence and theoretical reasons for expecting peripheral expansion to continue in most of the newer and even middle-aged urban regions. It grossly overestimates the likely "return to the city" of affluent Anglos and the exodus of the poor and of minority enclaves outward into suburbia. But to the extent that such patterns do develop, they will contribute to a strengthening of the private sector relative to the public in the central city, to its use as an adjunct of the outer metropolis, and to a weakening of the political potential of minority groups as their voting power is divided between the inner metropolis and the inlying suburban municipalities into which they penetrate.

INTO THE FUTURE:
A "HISPANIC RENAISSANCE" SCENARIO

The most likely force for a change in power relationships between the inner and outer metropolis, apart from the unforeseen, is the rapid and until recently unexpected growth of the Hispanic, predominanty Mexican-American/Chicano population and its continuing concentration within urban areas, most heavily in the Southwest.[4] The demographic pattern is already irreversible, with only the upper magnitude of this ethnic addition to our population remaining uncertain. What may follow from it is far less predictable. It seems clear that the old models of minority-majority relationships in U.S. urban areas, which apply so well to the development of the dual metropolis, will not hold with this new immigration. Four unusual factors lead to this conclusion. Two are almost certain: (1) a unique tie between the Hispanic minority and Mexico, which will condition Anglo policies toward the minority; and (2) a regional base for ethnic political power of an extent and character never before enjoyed by a minority. Two others are much more speculative: (3) development by this massive Hispanic minority in coalition with other minorities of a confrontation politics vis-à-vis the Anglo metropolis; and (4) with this a restructuring of certain facets of local public sector institutions to achieve a workable city for those who have been excluded from the mainstream of economic and civic involvement.

THE DEMOGRAPHIC DYNAMIC

Latinos/Hispanics now comprise the fastest-growing minority in the United States. Official estimates suggest an increase of almost 15% between 1973 and 1978 to a total of at least 12 million. Inclusion of undocumented aliens would probably bring the figure closer to 20 million. They will become the largest minority, exceeding Blacks in total numbers, before the turn of the century, perhaps even before 1990. They are largely an urban population. New York City has long had over 2 million Hispanics, over half of them Puerto Rican. Miami has over 200,000, mainly Cubaños. Rapidly growing Latino communities are evident in some of the cities of the Pacific Northwest, in Dever, and in some Midwestern urban centers, especially Chicago. But the Southwestern states remain the locale for the most rapid expansion of Latinos. There are over one and a half million in Los Angeles, primarily Mexican. California includes over 4 million Mexican-Americans and close to a million other Latinos by official count, plus another one and a half to one and three-quarters million undocumented aliens; and it is estimated that by 1990 nonwhite minorities will approximate 55 to 65% of the state's total population, the largest single segment being

Hispanics (Intern Research Project, 1978; Scott-Blair, 1978). A roughly equivalent number reside in Arizona, New Mexico, and Texas, 70% of them in the "Lone Star State."

Migration across the border with Mexico has been the largest single source of this population growth, with the great majority of the migrants originating in that country. The central driving force for this migration is the combination of rapidly increasing population and continuing mass poverty in Mexico, with little likelihood that these conditions will change substantially in the next few years (Fernandez, 1977; Lyons, 1977). Many factors facilitate and reinforce this flow. No oceans separate those already established in the United States from their relatives in Mexico; they draw both older and younger kin to this country. Despite the evidence of shifting family relationships in large urban centers (Maldonado, 1975), family values and ties remain strong among Latinos, especially those of Mexican heritage (Sotomayer, 1971; Valle and Mendoza, 1978). For many visiting back and forth between the U.S. Southwest and Mexico involves no more distance or expense than Anglos encounter going to see their kin elsewhere in the United States.

Moreover, many on both sides view the border only as a political and bureaucratic barrier established by force of arms for the economic benefit of the dominant United States (Fernandez, 1977). The Southwest was Mexico before the Treaty of Guadalupe Hidalgo in 1848. For a large proportion of the territory's Latinos, it has remained culturally Mexican in the intervening years. Both older action groups within the Mexican-American ambience and the contemporary Chicano movement share that orientation, with the latter asserting that its cultural homeland is Aztlán, the northernmost territories of the Aztec empire, which encompassed the Southwest and extended southward to Mexico City. Thus, many Latinos in the United States and many Mexican citizens privately view migration northward across the border as justified morally as well as economically, regardless of U.S. law.

Although substantial limitation of the massive flow of migrants across the border is theoretically possible, it would require a complex program of controls on the U.S. side, increased incentives for would-be migrants to remain in Mexico and other countries of origin, and extensive cooperation by the major sending nations to reduce migration to the United States (Huss and Wirken, 1977). The failure of the Carter administration and Congress to act expeditiously to establish and implement such a program is indicative of the immense difficulties entailed and obstacles that would have to be overcome. Those closest to the problem, whatever their preferences, do not anticipate achievement of more than partial measures, many of them hardly more than symbolic.

Finally, migration is not the only source of the increasing proportion of Latinos in the United States. The age structure and values of the inmigrants result in higher fertility rates than among Anglos and will continue to do so for at least a generation (Bradshaw and Bean, 1973). The maximum effects of this fertility on such factors as employment and voting power will become evident as today's young children and those to be born in the years ahead reach adulthood. But the impact on schools and other child-care systems is already evident.

THE "MEXICAN CONNECTION"

The ties between Mexican-Americans and Mexico involve more than culture, kinship, and ideology. They are politicized in ways which, taken together with migration and territorial proximity, create a unique situation which will have consequences for the role of Latinos in U.S. domestic politics and urban policies. Economic ties between the United States and Mexico have been increasing and will continue to do so. Petroleum is at the center of this commerce, but Mexico is also being viewed as a source of both electrical (geothermal and oil-fueled) and consumer goods production for importation into the United States. Employment of Mexican labor in U.S. agribusiness remains a magnet for both legal commuters and undocumented migrants. Focal points for these relationships as well as gateways for illegal migration are the "sister cities" along the border and the regions around them (Blair, 1977).

The leverage given to Mexico by U.S. concern for its economic interests, especially oil, is evident in the unusually strong terms in which President Portillo has expressed concern for Mexican nationals in the United States, both legal and undocumented. The appointment of an agency to be a Mexican presence on the U.S. side of the border to represent the rights of Mexicans is unprecedented. The treatment of Mexican citizens in this country is thus both a domestic issue for Mexican-Americans/Chicanos and a matter of international relations between the two nations. An increasing number of agencies, commissions and projects dealing with border problems and with the development of the border region, both bilateral and unilateral on the part of the United States, will have the same dual character politically. Adequate representation of Mexican-Americans will be expected. The policies and programs undertaken through such agencies and organizations will impact in a variety of ways, not only upon the urban areas throughout the U.S. Southwest, but also upon the Mexican side as well since the borderland's economy involves flows in both directions.

Although Latino groups tend to be focused primarily on local issues of discrimination and exclusion, the responsiveness of their leaders to the larger issues in which Mexico has an interest is evident and mutually advantageous. It fits in with ambitions for the rapid politicization of Hispanics, a formidable task for which every bit of help is needed. The connection has some minor antecedents: the 100,000-member League of United Latin American Citizens (LULACS) has enjoyed a 10 million dollar scholarship program sponsored by the Mexican government to enable Americans of Hispanic descent to study in Mexican universities. LULACS leaders have met several times with the president of Mexico. The more directly political, somewhat less middle-class Mexican-American Political Association (MAPA), rooted mainly in California, seeks Mexican celebrities to help them raise funds for organizing barrios throughout the state. Mexico in turn could very much benefit from a "Latino lobby" capable of expressing an interest in such matters as the continuing dominance of the U.S., which tends to be built into most bilateral economic development programs (Clement, 1978a, 1978b; Fernandez, 1977).

POLITIZATION AND REGIONAL POWER

Were it not for the rapidly changing character of U.S.-Mexican relationships and the build-up of sheer numbers of Hispanics, there would be little reason to see "brown power" as a major factor in local or regional, much less national, politics. Even where Latinos have comprised a substantial segment of the local citizenry, their role in decision-making has been at most modest and often less than that (Klapp and Padgett, 1960; D'Antonio and Form, 1965; McWilliams, 1968; Strauss, 1968). Community participation in programs and agencies of the "war on poverty" period may now be seen as a stepping-stone in Hispanic political development, but not much more (Kurtz, 1973). What is striking in the Los Angeles area today is not how much clout Mexican-Americans wield, but how little relative to their numbers. Part of the reason for the lack of political potency is poverty and racial discrimination and their by-products: Hispanics have the highest unemployment, the lowest income level, and the greatest drop-out rate from school. Many have brought with them from their home countries a mistrust of government and an even deeper suspicion of politics and politicians. They are more oriented toward family, neighborhood, and in some cases gangs than to political party or city hall, much less state or national capitals. Moreover, nationality distinctions among Latinos are strongly felt identities, and a common fate cannot be assumed to lead quickly and easily to common cause. Even within the Mexican/ Chicano communities there are serious ideological divisions and political rivalries.

Yet there is a growing expectation that this will change and that Latinos will increasingly be a force to be reckoned with in the eighties, especially in the cities and states where their numbers will be greatest (Time, 1978; Kirsch, 1978). One "futurist," seeking to assess the change process to the turn of the century, has sketched two scenarios in which extreme transformations would be produced (Downes, 1977). In one the United States would become by continuous but essentially peaceful increments a predominantly Hispanic society. In the other hostility would be maximized in the form of struggles between Blacks and Hispanic groups, on the one hand, and on an even larger scale (as the Hispanic population continued to burgeon) between an increasingly nationalistic and separatist Hispanic minority and a still dominant but increasingly repressive Anglo majority. The latter scenario foresees Mexico becoming active on behalf of a separatist movement in the Southwest. As Downes stipulates, neither of these sketches is a prediction; they serve to force attention to the dynamics already underway and to the range of possibilities. They also remind us that many aspects of a present state of affairs can be projected with confidence for only a few years ahead, that substantial shifts in the policies of major institutional systems or popular values can begin to cumulate rapidly within five years, and that revolutionary alterations can occur within two decades (Joseph, 1974). If there is to be a "Hispanic renaissance" in at least some of our urban centers, its emergence and direction should be evident within the coming decade.

To repeat, the nexus for politicization and regional power for Latinos is the unprecedented increase in numbers. Unlike the Blacks moving northward, Mexicans arrive in the urban centers of the Southwest (and elsewhere, of course) at a time when the ability of the Anglo power structure within the inner metropolis to contain and control a large new minority group has been weakened both by "white flight" and by policies of equalization and restitution for minorities as have been articulated and implemented. Many Latinos who have moved into local and state governmental bureaucracies feel safe enough, or worry little enough, to engage directly and extensively in eliciting and facilitating political participation in the barrios. It is almost a reverse of the old notion of using voting blocs to gain access to governmental jobs and contracts through patronage. Affirmative action programs however flawed and anticipatory accommodations to brown power by some Anglo political leaders, are creating beachheads in the public sector from which the political potential of the proliferating Latino population can be organized. Local and Mexican governmental pressures will combine throughout the coming decade to promote a U.S. policy which (under the guise of "amnesty" or whatever) converts residency into citizenship. In such circumstances the many legal aliens who have avoided citizenship because of identification with Mexico will change status.

If other pieces fall into place, the pace of Latino politicization and the consolidation of power in central cities will pick up rapidly through the latter part of the 1980s. The conservatism and assimilationism among many older, more established middle-class Latinos appears to be giving way to the concurrent experiences of militant ethnicity and recurrent Anglo resistance to often modest claims for political parity within the local community. At the other end of the socioeconomic spectrum the organization of migrant farm labor by Chavez not only makes explicit the industrial character of agribusiness, but also provides a basis for linking Mexican-Americans in the small town and farming areas in the peripheries of the still expanding urban regions and in the hinterlands beyond to the population and political centers of the Southwest. Mexican-American and Chicano political activists, with all their ideological and organizational differences, are in touch throughout the region.

A role in expediting politicization is also being played by an expanding cadre of artists and academics whose search for connections with the common culture of *La Raza* is evidenced by an efflorescence of street art and a commitment to Chicano and Mexican-American studies and research growing out of the historical experience and present condition of Mexicans on both sides of the border (Ornelas et al., 1975). They, too, constitute a regional network, and their contributions to a developing image of the Southwest as a "Third World" presence within the still dominant Anglo society is deliberately independent of conventional Anglo radicalism and its ideological use of Third World symbolisms. Bilingualism is a central element at both verbal and pictorial levels. Protecting and advancing it makes essential a forced draft build-up of influence in school districts and state legislatures. It is an issue of common cause for *artistas, politicos, profesorado, y el pueblo en general.*

COALITION AND CONFRONTATION

Even though the process of politicization among Hispanics in the Southwest may fall short of the ambitions of its more ardent advocates, there is no doubt that there will be a substantial increase in their representation in city, county, and state legislative and administrative bodies and in school districts throughout the 1980s. And in spite of internal conflict within the movement of *La Raza,* the inter-city and inter-state networks throughout the Southwest will thicken and expand. But one of the major uncertainties is the future relationship of Latino politics with the interests and politicking of other minorities in the region. As a separate force the Mexican Americans can have a good deal of power by the end of the 1980s. But were they and the other minorities to develop a coalition strategy and implement it consistently, they could become the dominant power in a number of major and middle-sized cities and be strong enough in state legislatures to secure much of what they would need to pursue a reconstruction of urban policies and practices in accord with their interests and ideological commonalities.

Chances for such coalitions seem best between Latinos and a growing population of Asian origin and heritages. Efforts at the development of a pan-Asian organization and influence are underway in a number of cities. The Filipinos whose culture now incorporates both an Asian origin and Hispanic influences, are a likely linkage. Common cause among these groups could emerge around the place of the family and of neighborhood informal networks in the restructuring of social services, the means for preserving linguistic and cultural traditions within a public education framework, and overcoming Anglo resistance to the changes necessary for achievement of some kind of parity in the occupation-income structure for nonwhite minorities. There also seems some likelihood that Native Americans, another growing segment of Southwestern urban populations, could be brought into such coalition politics. Pan-Indian organization has also been underway, even though the high priority given to tribal affiliation and identity makes it a reluctant step for some, taken only out of political necessity. Native Americans share with Mexican-Americans a long history in the Southwest, an understanding of Anglos as invaders and conquerors, and a confidence in the integrity of their cultures. Their symbolic role in local and regional politics is likely to extend far beyond that which numbers alone would make likely in the 1980s because of the vitality of tribal and cultural revival, the development of cadres of youthful advocates of Native American identity and activism, and the unique place and meaning of these peoples in U.S. history.

Any scenario for urban minority coalitions is invested with a multitude of uncertainties. But the greatest uncertainty is that attending the role to be played by Black communities in the Southwest. There is some history of antagonism between them and Latino groups, especially in cities where they have competed for occupational opportunity and residential territory. Expressions of prejudice against Blacks are common among such Latinos, although civic and political leaders understand that racial stereotyping of one another by minorities is an emulation of the Anglo society which can only contribute to its continuing dominance. If such understanding could lead to a coalition formation which included Blacks, especially where their numbers are significant, then political control of the central city by the minorities would be within reach.

Only with such control could consideration of a reconstruction of the public sector in the central city be more than an exercise of the political imagination. But challenging all the ways in which governmental policies and practices support the interests of the Anglo metropolis instead of those of the minorities and lower-income residents in general, would certainly bring a confrontation with Anglo power at the state as well as local levels. Such a course would seem both frightening and perhaps futile to Anglos, and it would be a clear indication that minority group leaders had overcome what Orans (1971) has called the "rank concession syndrome," at least in the political domain. The "syndrome" involves emulation of the society, strata, or group whose superior rank has been conceded. Beliefs, lifestyles, and practices of the superordinate group are openly adopted or disguised somewhat by adaptation by the minority. Stratification within the minority then reflects this emulation insofar as it results from individual mobility via success in the marketplace. Such a pattern reduces solidarity within the subordinate group and is conducive to successful members being co-opted in the political domain. The political pathway to improvement in rank, when it is undertaken independently, requires solidarity within the group for success and stimulates cultural solidarity between the elite within the minority and those of ordinary rank if the socioeconomic distance between the former and the latter has not already become too great.

The development of such solidarity within minority groups, the creation of coalitions among them in order to achieve control over major central cities, and the utilization of the municipal governments in a coordinated way throughout a region to confront the dominance of the Anglo segment of that region—such a scenario surely seems improbable. But twenty years ago few could foresee the evolution of the dual metropolis, the turbulence of the sixties and the early seventies, or the emergence of Hispanics as the largest U.S. minority and concurrently the "Mexicanization" of the Southwest. This is not to assert that the scenario for a "Hispanic renaissance" should be viewed as a good prediction, but that we should be cautious about our ability to know what will not happen.

EL RENACIMIENTO HISPANOAMERICANO: LA CIUDAD MODESTA

The concluding segment for such a scenario must be even more speculative. It involves trying to answer the question: What kind of urban order would a coalition of minorities within the inner city seek as part of the confrontation politics stimulated by continuation of tax/spending limitations imposed by the dominant class at the state and federal levels?[5] A conditioning assumption is that the municipalities which they would come to control would show the effects of Anglo retrenchment. Indeed, the new leaders and managers of the inner metropolis would be able to engage in a politics of confrontation only if it were accepted that they had little to lose and much to gain, and that it was their historic fate to take high risks on behalf of a new American pluralism. Finally, their ambition would be for a much more modest city than has satisfied the Anglo upper middle class and elite strata with their addiction to ever more grandiose facilities for expensive shopping and leisure pursuits. For a growing proportion of residents of the inner metropolis the comparisons to be drawn will be with the Third World cities they have left behind, or the towns and rural areas of the remaining underdeveloped hinterlands within the United States which they, too, have abandoned. What follows, then, are not predictions but sketches of possibilities. They are potentialities produced by the concurrence of a policy of retrenchment of urban institutions in the public sector at the expense of minorities and poor whites encapsulated within the central city, and of a massive influx of new immigrants, mostly non-Anglo, mainly Hispanic, with a majority from Mexico.

The economy. The city's economy will begin to display a different mix of private, quasi-public, and public enterprise. A great many of the new immigrants are would-be petty bourgeois: small producers, merchandisers, and service providers capitalizing on an extended family whose members can be both owners and employees. Much of what constitutes the periphery sector of the economy is their "natural" milieu as both sellers and buyers. Its facilitation, protection, and expansion will be municipal policy, for which support will be sought in state capitals and in Washington. Collectives, cooperatives, and local development corporations will receive new interest and support, but only where they do not impinge seriously on the investments and markets of the new petty bourgeoisie. Research oriented toward the concerns of the new controlling coalition will challenge the real value to the central city of enterprises which appear to serve primarily the interests of the outer metropolis or supralocal corporations.

Municipal budgets and services. Until the Anglo majority in the larger society would deem it in their interests to reverse retrenchment, it is likely that the new central-city regimes could only reallocate their limited budgets to serve as well as possible the lower- and middle-income majority of the municipality's residents. Such a policy could also be part of a strategy to bring pressure on the more affluent strata, both within the city and in the outlying Anglo metropolis. One possibility is the assignment of curtailed police and firefighting personnel and equipment on the basis of the incidence of crimes and conflagrations, greatly increasing service to older and lower-income areas. Private protection agencies of the sort already employed by some shopping centers and condominiums would be encouraged for more affluent residential and business areas whose owners believed it necessary to compensate for the reduction or withdrawal of municipal protective services. Use taxes and service charges levied differentially by zones set to reflect user income are another possibility. Income taxes on earnings within the city regardless of residence could receive increasingly favorable attention, as might licensing for land use related to proportionate use by residents and nonresidents, whether office buildings or centers for the performing arts. But care would be taken not to imperil labor-intensive service industries providing jobs for minority group members, such as hotels, restaurants, and night clubs. A good many municipal regulatory agencies (which appear to have been extended at substantial cost to benefit Anglo businesses and unions, as in the construction industry) might well be simplified, curtailed, and perhaps decentralized.

Social services. A major concern for the new inner-city leaders and managers must be to find cost-effective approaches to welfare and social service provision which overcome present tendencies to cultural insensitivity, overbureaucratization, encouragement to dependency, and weakening of familial ties. Innovations would be likely linking support and service provision to the "natural systems" generated within minority group neighborhoods and family networks, an approach on which attention is being focused within both Latino and Asian community groups. Here decentralization and reduction of professional bureaucracies are essential elements of change.

Public education. Keeping youth in schools and the successful teaching of marketable skills along with the "basics" will have priority in the minority communities over educational doctrines and technologies created by an Anglo professional establishment which has become increasingly oriented toward the lifestyles and occupational expectations of the outer metropolis. Emphasis will be on strong teachers, effective discipline, bilingualism and a credible connection with community needs. Concerns for cutting costs will lead to challenges of both existing educational bureaucracies and Anglo-dominated teachers' unions. Cultural pluralism and residential segregation indicate a shift away from busing for desegregation in at least the earlier years, plus substantial decentralization of curriculum planning, textbook acquisition, and even staffing.

AND THEN....

The changes just sketched do not produce a stable urban system, one which would "settle down" for a few generations. They would only alter the character and distribution of conditions treated as problematic. But neither can the present patterns be held stable for very long in the face of the pressures generated by immigration, internal migration, segregation, and chronic inequality distributed by race and residence.

NOTES

1. This is not the appropriate place for an extended methodological discussion. The approach follows what Kaplan (1964) has called the *pattern model* of explanation, as against the *deductive model,* which is embodied in most of the increasingly sophisticated statistical analyses of urban data. One model is not better or more scientific than the other. Deductive explanations are always more complete and precise, but their givens preclude the complexities and two-way, multiple relationships which characterize our still evolving urban systems. "Rather than saying that we understand something when we have an explanation for it, the pattern model says that we have an explanation for something when we understand it.... Not all explanation consists in fitting something into a pattern already given. The task of explanation is often to find or create a suitable pattern" (Kaplan, 1964: 335, 336).

2. There is no necessary connection between intentional lifestyle subcultures and the private sector, although one might argue that support of such leisure and consumer patterns through a socialist economy would likely be constraining of some innovations. Certainly some socialist governments have been far more supportive of ethnic subcultures than has U.S. capitalism, and theoretically similar support could be extended to lifestyle variations with other bases. But this would reduce, if not eliminate, the present function of such subcultures as expressions of class level as well as self-image and consumer ideology. One can only speculate on whether such innovativeness would persist, or to what degree, apart from status competiton.

3. Allman indicates that he draws heavily on the work of Professor Franz Schurmann of the University of California at Berkeley, and that work Allman did for the Pacific News Service City Project also feeds into his analysis. One assumes that his various sources are not responsible for Allman's convenient use of statistics, in which relatively short-term changes in such indicators as unemployment (down), federal aid (up), and decline of Black ghetto populations are used to demonstrate the passing of the "urban crisis" in cities such as Cleveland, Ohio.

4. Ethnic, class, regional, and ideological differences within a minority which is in the process of being mobilized, at least to some degree, through social movements and political activity often become associated with terminological preferences with respect to group identification. An outsider is especially subject to being charged with ignorance and/or insensitivity by members of the minority who consider the terminology used to be inappropriate, a charge which may often have some substance. Hispanic and Latino are both used to designate all Spanish-speaking immigrant groups from Latin-American nations. Chicano usually designates second- or third-generation Mexican-Americans (Chicano y Chicana would be less sexist), but some of them apparently prefer to continue the identification term, Mexican-American. "The Chicano" clearly is an outgrowth of the experience in the United States of immigrants from Mexico, often carries with it political connotations, and may increasingly be extended to include first-generation inmigrants, especially youth, who merge with their second-generation peers culturally and politically.

5. In one sense it is inappropriate for an Anglo even to attempt that speculation. Such a stricture would have special force if what follows were offered as

advice to minority groups as to what they "ought" to do. It is not. Rather, it is an effort to extrapolate into an imagined new political context presently visible strategy inclinations and policy predispositions generated within the emerging leaderships of Latino, Asian, and Native American minority groups in the Southwest, as I have been able to perceive them through both public statements and actions and data from informants within those groups.

REFERENCES

ALLMAN, T. D. (1978) "The urban crisis leaves town." Harper's 257 (December): 41-56.
AVERITT, R. T. (1968) The Dual Economy: The Dynamics of American Industry Structure. New York: Horton.
BECK, E. M., P. M. HORAN, and C. M. TOLBERT (1978) "Stratification in a dual economy: a sectoral model of earnings determination." Amer. Soc. Rev. 43 (October): 704-720.
BERRY, J.L.B. and J. D. KASARDA (1977) Contemporary Urban Ecology. New York: Macmillan.
BLAIR, C. P. (1977) Prepared statement. Recent Developments in Mexico and their economic implications for the United States. Hearings before the Subcommittee on Inter-American Economic Relationships of the Joint Economic Committee, U.S. Congress, January, 1977. Washington, DC: U.S. Government Printing Office.
BLAUSTEIN, A. F. (1978) "Proposition 13 = Catch 22." Harper's 257 (November): 18-22.
BLUESTONE, B., W. M. MURPHY, and M. STEVENSON (1973) Low Wages and the Working Poor. Ann Arbor, MI: Institute of Labor and Industrial Relations, University of Michigan.
BRADSHAW, B. S. and F. BEAN (1973) Trends in Fertility of Mexican-Americans, 1950-1970. Social Sci. Q. (March): 688-696.
BRIMMER, F. A. (1969) "The Negro in the American economy," in J. F. Kain (ed.) Race and Poverty. Englewood Cliffs, NJ: Prentice-Hall.
Bureau of Labor Statistics (1978) Consumer Expenditure Survey: Integrated Diary and Interview Survey Data, 1972-73 (Bulletin 1992). Washington, DC: U.S. Government Printing Office.
——— (1964) Consumer Expenditures and Income: Survey of Consumer Expenditures, 1960-61 (Report 237-38). Washington, DC: U.S. Government Printing Office.
Bureau of the Census (1978) Current Population Reports: Consumer Income. Washington, DC: U.S. Government Printing Office.
CLEMENT, N. (1978a) Binational Coordination and Planning on the California-Baja California Border. San Diego: Border States University Consortium for Latin America, San Diego State University.
——— (1978b) United States-Mexico Economic Relations: The Role of California.

San Diego: Center for Research in Economic Development, San Diego State University.
COLENUTT, R. J. (1972) "Do alternatives exist for central cities?" in H. M. Rose (ed.) Geography of the Ghetto: Perceptions, Problems, and Alternatives. DeKalb, IL: Northern Illinois University Press.
D' ANTONIO, W. and W. H. FORM (1965) Influentials in Two Border Cities: A Study in Community Decision-Making. Notre Dame, IN: University of Notre Dame Press.
DAVIS, K. (1955) "The origin and growth of urbanization in the world." Amer. J. of Sociology 60 (March): 429-437.
DOBRINER, W. (1963) Class in Suburbia. Englewood Cliffs, NJ: Prentice-Hall.
DOUGLAS, H. (1925) The Suburban Trend. New York: Century Co.
DOWNES, R. (1977) "The future consequences of illegal immigration," Futurist 11 (April): 125-127.
DOWNS, A. (1968) "Alternative futures for the American ghetto." Daedalus 97: 1331-1378.
FERNANDEZ, R. A. (1977) The United States-Mexican Border: A Political-Economic Profile. Notre Dame, IN: Notre Dame University Press.
GALLUP, G. (1978a) The Gallup Poll. San Diego Union (August 27): A-27.
--- (1978b) The Gallup Poll. San Diego Union (August 28): A-12.
GREER, S. (1962a) The Emerging City. New York: Free Press.
--- (1962b) Governing the Metropolis. New York: John Wiley.
HARRIS, C.D. (1943) "Suburbia." Amer. J. of Sociology 49 (July): 1-13.
HUSS, J. D. and M. J. WIRKEN (1977) "Illegal immigration: the hidden population bomb." Futurist 11 (April): 114-120.
Intern Research Project (1978) Third World Population in California. Sacramento. CA: Council on Intergroup Relations, Office of the Lieutenant Governor (revised February).
JOSEPH, E. C. (1974) "What is future time?" Futurist 8 (August): 178.
KAIN, J. F. (1968) "Housing segregation, Negro employment, and metropolitan decentralization." Q. J. of Economics 82: 175-197.
KAPLAN, A. (1964) The Conduct of Inquiry. San Francisco: Chandler.
KASARDA, J. D. (1972) "The impact of suburban population growth on central city service functions." Amer. J. of Sociology 77, 6: 1111-1124.
KIRSCH, J. (1978) "Chicano power." New West 3 (September 11): 235-240.
KLAPP, O. E. and L. V. PADGETT (1960) "Power structure and decision-making in a Mexican border city." Amer. J. of Sociology 65, 4: 400-406.
KOTTER, P. J. and P. R. LAWRENCE (1974) Mayors In Action. New York: John Wiley.
KURTZ, D. V. (1973) The Politics of a Poverty Habitat. Cambridge, MA: Ballinger.
LIEBERT, R. J. (1976) Disintegration and Political Action: The Changing Functions of City Governments in America. New York: Academic Press.
LONG, N. E. (1958) "The local community as an ecology of games." Amer. J. of Sociology 64 (November): 251-261.
LYONS, G. (1977) "Inside the volcano." Harper's 254 (June): 41-55.
McWILLIAMS, Carey (1968) North From Mexico: The Spanish-Speaking People of the United States. New York: Greenwood.

MALDONADO, D. (1975) "The Chicano aged." Social Work 20, 3: 213-216.
MARCH, M. S. (1968) "The neighborhood center concept." Public Welfare 26: 97-111.
MARR, P. D. (1972) "Functional and spatial innovation in the delivery of governmental social services," in H. Rose (ed.) Geography of the Ghetto: Perceptions, Problems, and Alternatives. DeKalb, IL: Northern Illinois University Press.
MULVIHILL, D. J., M. M. TUMIN and L. A. CURTIS (1969) Crimes of Violence. Staff Report to the National Commission on the Causes and Prevention of Violence, Vol. 2. Washington, DC: U.S. Government Printing Office.
ORANS, M. (1971) "Caste and race conflict in a cross-cultural perspective," pp. 83-150 in P. Orleans and W. R. Ellis, Jr. (eds.) Race, Change, and Urban Society. Beverly Hills: Sage Publications.
ORNELAS, C., C. B. RAMIREZ, and F. V. PADILLA (1975) Decolonizing the Interpretation of the Chicano Political Experience. Los Angeles: Chicano Studies Center Publications.
ROSE, H. (1978) Black Suburbanization. Lexington, MA: Lexington.
SCHNORE, L. (1972) Class and Race in Cities and Suburbs. Chicago: Markham.
––– (1957) "Metropolitan growth and decentralization." Amer. J. of Sociology 63 (September): 171-180.
SCOTT-BLAIR, M. (1978) "1990 ethnic majority report stirs debate." San Diego Union (July 16): A-3.
SENNETT, R. (1970) The Uses of Disorder: Personal Identity and City Life. New York: Vintage.
SMITH, J. (1970) "Another look at socioeconomic status distributions in urbanized areas." Urban Affairs Q. 5: 423-453.
SOTOMAYOR, M. (1971) "Mexican American interaction with social systems." Social Casework 52, 5: 316-324.
STRAUSS, M. P. (1968) "The Mexican-American in El Paso politics," in C. J. Wingfield (ed.) Urbanization in the Southwest. Public Affairs Series No. 1. University of Texas at El Paso.
TAYLOR, G. R. (1915) Satellite Cities. New York: Appleton-Century-Crofts.
Time (1978) "It's your turn in the sun." Time 112 (October 16): 48-61.
TURNER, J. (1970) "Blacks in cities: land and self-determination." Black Scholar 1:9-20.
VALLE, R. and L. MENDOZA (1978) The Elder Latino. San Diego: Center on Aging, San Diego State University.
WALTON, J. (1968) "Differential patterns of community power structure: an explanation based on interdependence," in T. N. Clark (ed.) Community Structure and Decision-Making. San Francisco: Chandler.
WINTER, W. O. (1969) The Urban Polity. New York: Dodd, Mead.

12

Urban Education: Complex Problems— No Simple Solutions

CHAVA NACHMIAS

EDUCATION AND SOCIAL MOBILITY

☐ THE NOTION OF SOCIAL EQUALITY inspired by the Horatio Alger "log cabin to president" success story, is a cherished, though elusive concept in the open-class ideology of American society. Differentiation in power, prestige, and wealth are viewed as legitimate rewards for those who have talent and ambition to use the opportunities for upward mobility which are allegedly available to all who want it.

The system of stratification in America involves several aspects including income, type of residence, ethnicity, race, occupation, and other related factors. Occupation, however, is perhaps the most important component of the American stratification structure.

> The occupational structure in modern industrial society not only constitutes an important foundation for the main dimensions of social stratification, but also serves as the connecting link between different institutions and spheres of social life and therein lies its great significance. The hierarchy of prestige strata and the hierarchy of economic classes have their roots in the occupational structure. [Blau and Duncan, 1967: 167]

Urban America, like other complex industrial societies is characterized by an elaborate system of occupations; occupational positions, requiring extensive amounts of skill, training, and responsibility are rewarded with a disproportionate amount of wealth, prestige, and power. The study of social mobility is concerned with the determinants of individual movements in and out of these occupational positions that are so closely linked with one's relative position in the stratification system. Much attention has been given to intergenerational mobility, describing the process by which children move to positions in the stratification structure that are lower or higher than the position held by their parents. Examination of occupational trends in the last several decades reveals a high rate of intergenerational mobility (Lipset and Zetterberg, 1956; Blau and Duncan, 1967). These patterns can be explained in part as a function of the changing occupational structure in American society. The most significant labor force development of the postwar period has been a strong upward trend in total employment. During the last twenty-five years, total employment has almost always increased above the previous year's level (Sorkin, 1974). Concurrent with this trend, there has been substantial increase in the proportion of professional and clerical workers and a decline in the percentage of workers in semiskilled and unskilled occupations. These changes in the occupational structure have been associated with rising educational requirements within the labor force. A major outcome of this has been to make education the main avenue for mobility. This, in turn, has created a widespread belief in the educational system as the dominant means by which the large majority of rewarded positions are attained. Social status is increasingly achieved, rather than simply ascribed; and achieved, moreover, by means of educational attainment.

In their comprehensive study of *The American Occupational Structure*, Blau and Duncan (1967) found that while there is a considerable effect of ascription on occupational attainment, it is largely mediated through educational attainment. Using data from a 1962 national sample of males twenty to sixty-four years old, the authors constructed a causal model of occupational attainment using father's educational and occupational status, as the determinant of the son's education, his first job, and his occupation, in 1962. Using a series of recursive equations, they estimated the relative significance of social origin and educational attainment on the individual's first job, and current occupation. Moreover, when social origin was statistically controlled for, educational attainment was found to be more significant in accounting for later occupational status than the first job. Blau and Duncan's work points to the crucial role that education plays in the status attainment process. One of the effects of this has been the realization that higher education has become the main avenue for mobility; this has also put a burden upon the secondary school, where the decision to go to college is made, and where the curriculum prerequisites for entrance are obtained. Thus, the decision and performance in the eighth to twelfth grade became the most crucial determinants of future status.

THE STATUS ATTAINMENT MODEL

Given the emphasis on education, the focus of later mobility studies has been upon access to educational opportunities, upon the quality and variability of education made available, upon the cognitive skills required for educational attainment, and upon the motivations and aspirations of individuals to utilize available educational resources. The questions to which these studies address themselves to are: To what degree is an individual's achievement dependent on factors other than his own ability aspiration and efforts? What are the organizational and social psychological mechanisms of this dependence? And to what extent is ability, aspiration, and effort themselves dependent on factors other than the individual's own experiences and prior achievements?

A child's educational performance is dependent upon the extent to which his experience outside of school has prepared him to play the student role. Social background shapes the individuals' ability to some degree, which, in turn, affects their academic expectations, their performance, and their eventual occupational attainment and earnings.

Among the most significant attempts to describe the process of occupational attainment via educational achievement is research on "status attainment." This research, originally developed by a group of researchers affiliated with the University of Wisconsin (it became known as the "Wisconsin Model"), has been identified with the sociopsychological approach to social mobility. The major contribution of this approach has been to call attention to interpersonal factors such as significant-others influence as a source of status aspiration and status attainment. Occupational attainment and earnings are estimated in a recursive path model, which describes the process through which the

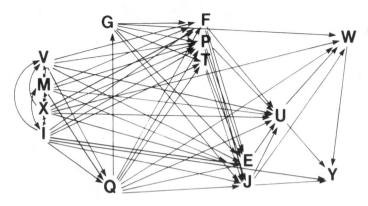

FIGURE 3: THE STATUS ATTAINMENT MODEL

VARIABLES: V=father's education, M=mother's education, X=father's occupational status, I=parents' average income, Q=mental ability, G=high school grades, T=teacher's encouragement, P=parents' encouragement, F=friends' plans, E=educational aspirations, J=occupational aspirations, U=educational attainment, W=occupational status, Y=earnings.

NOTE: The curved two-headed arrow represents correlations for which no causal interpretation is offered. The straight unidirectional arrow represents direct causal effect.

SOURCE: Sewell and Hauser, 1975.

effects of social origin and ability are mediated by a number of sociopsychological factors and academic performance. The model is presented in Figure 3.

The model starts by positing the dependence of mental ability (Q) on socioeconomic background (V, X, and I). High school grades (G) depend both on socioeconomic background and on ability, but only ability turns out to affect grades directly. Socioeconomic background, ability, and grades are next assumed to affect three perceptions of social support from significant others: teachers' encouragement to attend college (T); parents' encouragement to attend college (P); and the perception that most of the youth's friends plan to attend college (F). Educational aspirations (E) and occupational aspirations (J) are assumed to depend on all of the preceding variables, background, ability, grades, and perceived encouragement. Finally, educational attainment (U) is affected by most of the preceding variables and occupational aspiration, as well as by father's occupational status and the individual's educational attainment. Earnings (Y) are directly affected only by the individual's occupational aspiration and educational and occupational attainment.

In summary, the model postulates that social origin affects mental ability, that both social origin and ability affect significant others' support, and that all of these account for aspirations, educational attainment, and occupational status and earnings achieved by the individual.

This analysis of the model of socioeconomic achievement reflects the complexity of the process and highlights the interaction between ascription and achievement as determinants of status. Sewell and Hauser in summarizing their study make the following statement:

> Our finding of the distinct mediating and direct effects of ability, schooling and occupational status adds to the preponderance of evidence that factors of achievement as well as of social background enter importantly in the processes by which social goods are distributed in the United States. However sophisticated our notions about social origins, it is not possible to give an accounting of the distribution of education, occupation, and income in the United States that excludes individual achievement and ability. [Sewell and Hauser, 1975: 183-184]

While the model of status attainment has been originally used to describe data on a subsample of farm residents, it has subsequently been applied to small, medium, and large cities. In each case, the original model was supported with only minimal modifications. The great majority of that research based on the status attainment model has either investigated white students or failed to make systematic black-white comparisons. Recently however, a few studies have reported racial differences (Carter et al., 1972; Hout and Morgan, 1975; Kerckhoff and Campbell, 1977; Portes and Wilson, 1976). These studies show that it is not possible to explain as much statistical variance in educational attainment of blacks as it is in whites. "Those variables which form a cohesive and fairly powerful model for whites are much less effective for blacks; early stages in the attainment process consistently explain later one more fully for whites than blacks" (Kerckhoff and Campbell, 1977).

In addition, socioeconomic status has a much more significant influence on educational attainment for whites than for blacks. Among blacks, there is generally a lack of continuity between social origin and attainment, and also a weak link between earlier school performance and later (high school) performance. The crucial factors in the attainment process of blacks, however, is their high school performance. If they perform well in high school, they are at least as likely as whites to obtain further education. Another point of difference is the greater relative importance of nonacademic experience (e.g., discipline) among blacks in predicting future attainment. Kerckoff and Campbell (1977) report that for blacks, discipline is a more effective source of prediction of attainment than grades in junior high school, whereas for whites, the reverse is true.

Despite the differences observed between the races, there are consistent similarities as far as the effects of IQ on either academic performance or on educational attainment. Thus, while the effect of social origin on attainment is not clear for blacks, cognitive skills are as important for blacks as they are for whites.

EQUALITY OF EDUCATIONAL OPPORTUNITY: THE FEDERAL ROLE

The crucial role of education for broadening opportunities for productive and rewarding participation in the affairs of society has raised throughout the nation's history the quest to equalize access to education of high quality. Spurred on by the civil rights movement of the 1950s and 1960s, equal educational opportunity became a top national policy priority. It was viewed by academics, the attentive public, and policy-makers as the base for all rights and responsibilities of active membership in society. The effects of education on equality of opportunity is a complex and perhaps one of the most controversial and pressing issues in America's cities.

The most significant question facing educators and policy-makers during the last decade has been: Could better or different schools create more equality of opportunity? More specifically, has education provided opportunities for mobility for the urban poor or the black population? Or, as has been claimed by many critics, have schools reinforced existing inequalities (Hurn, 1978)?

Historically, the desire to equalize educational opportunities has been advanced in terms of individual liberty. Accordingly, the formulation and the assessment of educational policies have been centered around school inputs, including physical facilities of schools, training of teachers, teacher-pupil ratios, libraries, laboratories, and racial integration. In the 1960s the definition of equality of educational opportunity was radically broadened to include school outputs. Operationally this has meant the results of tests of academic achievement. Viewed from this perspective, there are large inequalities of educational achievement between ethnic, racial, and class groups. Educational inequalities automatically preclude admission of low achievers to most institutions of higher learning, and as a result, their occupational opportunities are severely limited.

The research of "status attainment has had some significant implications for policies designed to reduce inequality of educational opportunity especially for blacks and the urban poor. The model of status attainment points to three important sources influencing directly and indirectly the level of educational attainment: socioeconomic status, cognitive abilities, and significant others. Thus, educational programs directed at either one or all of these sources would be expected to increase individual attainment. At the cognitive level, educational programs designed to increase the academic ability and performance of black and lower-class children, especially in the inner city, had to be considered. There is accumulating evidence that inner-city children enter school with an academic deficiency which tends to increase with more years of schooling. As the Coleman report has pointed out, by the time they have reached the twelfth grade, black children are well behind in their academic skills and achievement (Coleman et al., 1966). Similarily, it was suggested that lower-class children enter the school situation so poorly prepared to produce what the school demands, that failing is almost inevitable, and the school experience becomes negatively rather than positively reinforced.

Among the suggested strategies to improve cognitive skills are educational programs directed at the preschool level with intensive verbal and symbolic experiences, the postponement of the introduction of formal school subjects at the primary grades, to help build up a repertoire of verbal experiences—both speaking and listening (Goldberg, 1963).

The model of status attainment suggests, also, that programs designed to influence the "significant other of disadvantaged student would have an effect on their educational attainment" (Sewell, 1971). This would include teachers, peers, and parents. Positive and high standards of academic expectations conveyed by teachers and counselors are essential to raising self-esteem levels and to increasing academic motivation of disadvantaged urban students. Similarly, peers could be used to stimulate educational aspirations and achievement. The Coleman report has suggested that educational background and aspirations of other students in the school are strongly related to the school attainment of the poor and the black in the city. Factors associated with student body characteristics, such as proportions of students planning to attend college, can be easily controlled by busing students. The environment provided by the student body is especially significant for blacks who are most sensitive to the school characteristics. Since home background tends to influence black students less than white, one might expect the school to have more of a chance to affect the educational attainment of blacks.

Among other forms of intervention designed to influence significant others, it has recently been suggested that a combination of material inducement with a reward structure emphasizing peer group attainment is an effective strategy for motivating lower-class adolescents (Spilerman, 1971).

To equalize educational opportunities and presumably broaden the structure of occupational opportunities for low-income and minority groups, the federal government took initiative in activating various educational programs from the preschool level (e.g., Headstart) to college. Many of these programs embodied the "compensatory approach" to education and were primarily aimed at increasing the cognitive skills and the educational attainments of children from low-income backgrounds. Concomitantly, the federal government took the initiative of requiring school districts to racially integrate their schools. This policy was in congruence with the Coleman Report that found schools to make relatively little difference except as a place where students learn from each other. The following section describes and evaluates some of the major programs initiated by the federal government.

FEDERAL EDUCATIONAL PROGRAMS

The Elementary and Secondary Education Act (ESEA) has significantly altered the federal role in education, in particular in cities. Prior to 1965 the overall contribution of the federal government to education was quite limited. General federal aid to education bills were unable to win congressional approval, and federal programs were marginal even in their intended impact. The federal inroads into public urban education was to a large extent stimulated by the inability of locally based and funded public primary and secondary schools to equalize educational opportunities to all citizens. ESEA doubled federal funds for education in a single year and is at present the largest federal aid to education program. The main thrust of ESEA has been in "poverty-impacted" schools, educational training and research, and instructional materials. The act provided the following:

Title I: Financial assistance to "local educational agencies serving areas with concentration of children from low-income families" for programs "which contribute particularly to meeting the special needs of educationally deprived children." Grants would be made on application to the Office of Education on the basis of the number of children from poverty striken families.

Title II: Grants to "public and private elementary schools and secondary schools" for the acquisition of school library resources, textbooks, and other instructional materials.

Title III: Grants for schools for "supplementary educational centers and services" including remedial programs, counseling, adult education, specialized instruction and equipment, and the like.

Title IV: Grants to universities, colleges, or other nonprofit organizations for research or demonstration projects in education.

Title V: Grants to stimulate and strengthen state educational agencies.

In addition to ESEA some twenty separate federal educational and training programs for low-income populations were either activated or expanded during the mid-1960s and early 1970s. Table 28 lists the major programs according to the particular target group that the program was designed to serve and their appropriations for selective years. These programs can be further categorized according to the target variable(s) that the policy-makers intended to influence: programs developed to improve basic cognitive skills (e.g., Title I, ESEA) and those designed to increase educational attainment (e.g., Upward Bound, Educational Opportunity Grants, Talent Search). The success of the first types of programs can be assessed by higher test scores and their relationship to earnings. A measure of success for the second type of

TABLE 28: MAJOR EDUCATIONAL AND TRAINING PROGRAMS FOR LOW-INCOME PERSONS FUNDED AT FEDERAL LEVEL: 1969-1973

Program	Appropriations (in million dollars)	
	1969	1973
Elementary-Secondary		
ESEA of 1965		
Title I—educationally deprived	1123	1810[a]
Title VII—bilingual education	7	50[a]
Title VIII—dropouts	5	8[a]
Upward bound	(included in Talent Search)	
School lunch programs	162	226[a]
School breakfast programs	3.5	18[a]
Emergency school aid act	—	271[a]
Vocational education	248	434[a]
Teachers corps	21	38[a]
Young adults		
Neighborhood Youth Corps	301	41[a]
Job corps	280	193
Higher education		
Higher Education Act of 1965		
Title I—matching grants	9.5	15[a]
Title III—developing inst.	30	52[a]
Title IV		
Educational Opportunity Grants	134	274
Guaranteed loans	71	240[a]
Work-study	146	274[a]
Talent search	34	72[a]

[a]Expenditures
SOURCE: Levin, 1978, p. 529

programs is additional years of schooling acquired by individuals participating in the programs in comparison to their nonparticipating counterparts; more schooling, in turn, could be expected to be associated with higher annual earning.[1]

These two types of educational programs are critical in terms of the status attainment process. A significant increase in cognitive skills is expected to enhance academic performance and to raise the level of educational and occupational aspirations, which, in turn, will increase educational attainment and ultimately occupational attainment.

With respect to cognitive skills, the major thrust of the various programs has been on reading and arithmetic proficiencies as well as other cognitive components. Overall it appears that cognitive skills programs targeted at low-income urban populations were of very limited effectiveness. There is little systematic evidence of significant improvement in cognitive skills resulting from the programs, and the evidence suggests that improvements in cognitive skills do not significantly enhance occupational attainment and/or earnings. For example, Title I of the ESEA required evaluations by the local school districts and state educational departments, and the U.S. Office of Education assessed the effectiveness of the programs for the 1966-1967 and 1967-1968 school years. On the basis of test scores, the study concluded that "a child who participated in a Title I project had only a 19 percent chance of a significant achievement gain, and a 13 percent chance of a significant achievement loss, and a 68 percent chance of no change at all (relative to the national norms)" (Picariello, 1969). Furthermore, of some 1,750 projects that were identified as appearing to meet the criteria of success in improving the cognitive skills of disadvantaged children, only 41 were found to be successful when evaluated in systematic ways. Other evaluation findings of Title I have reached similar conclusions. Indeed, these programs shared some inherent limitations, and their implementation varied widely. Furthermore, as Gray pointed out, "an effective early intervention program for a preschool child, be it ever so good, cannot possibly be viewed as a form of inoculation whereby the child is immunized forever afterward to the effects of an inadequate home and a school inappropriate to his needs" (Gray, 1974). In fact, the program follow-through was developed to ascertain how to reinforce and possibly increase the cognitive gains of preschool children during the first years of elementary school. Preliminary findings suggest that the program and its various projects exert very few positive effects (Stebbins et. al., 1978)

From a policy perspective, findings focusing on the relationship between cognitive skills and earnings among low-income populations have also been discouraging. In general, the evidence suggests that the measurable impact of cognitive skills on earnings is relatively small. At most, some 10% of the variance in earnings or income can be attributed to test-score for age-specific populations. In a recent comprehensive evaluation it was found that only among white males was there evidence of a small and statistically significant effect of reading competence on earnings. For black and white females no apparent relation between reading proficiency and earnings was found.

The educational programs designed to increase the educational attainment of the disadvantaged constitute the second major strategy of government in attempting to equalize educational and occupational opportunities. The underlying assumption beyond the educational attainment programs is that greater equality in the distribution of schooling will inevitably lead to greater equality in the distribution of income. This assumption does, in fact, make an explicit causal link in educational attainment models, as discussed earlier.

Evidence concerning the effectiveness of the various governmental programs on the educational attainment of poor and minorities are rather inconclusive. Ribich and Murphy (1975), for example, analyzed a sample of students who had been studied earlier in a national high school survey, Project Talent. They found that students who attended high school in school districts characterized by higher expenditures tended to obtain slightly more schooling than students from lower-expenditures districts. In other words, compensatory education tends to increase the number of years that disadvantaged students might attend school. Relatedly, the estimated final educational attainments of Upper Bound students suggest that some 70% of them attended college, and almost one-third completed four years of college. However, evaluations of other programs produced less encouraging results. A national evaluation of the Neighborhood Youth Corps (NYC) program—designed to increase high school graduation rates of potential dropouts by providing paid work experience to high school students who are likely to drop out—found it to be unsuccessful. The probability of graduating from high school did not increase as a result of participation in the program. Similarly, an evaluation of NYC projects in Detroit and Cincinnati found similar results, including no reduction in dropouts, no increase in educational aspirations, and no improvements in scholastic achievements. Furthermore, an evaluation of the Basic Educational Opportunity Grants (BEOG) suggests that the effects of the grants on improving the educational attainments of low-income students is severely limited. BEOG is likely to provide relatively more aid to relatively well-to-do families and less to poor ones (Hansen and Lampam, 1974).

SCHOOL INTEGRATION

The evaluation of the effects of school integration is a formidable task. The Coleman Report was based on data collected before concerned attempts were made to desegregate schools; therefore, the report does not provide information on the effectiveness of integration for increasing the cognitive skills and the educational attainments of disadvantaged children. More recent research is far from being conclusive, although the ideological controversy over its implication might suggest the opposite. Some findings indicate that while achievement remains stable in white majority schools, it may become significantly lower in majority black schools. Nancy St. John (1975), for example, found that of the thirty-seven studies of black academic achievement and racial composition only seven report clear positive results on all tests of school achievement; eighteen report mixed results on different tests, and the rest either no difference or negative results. Thus,

> Desegregation has rarely lowered and sometimes raised the scores of black children. Improvements have often been reported in early grade, in arithmetic, and in schools over 50 percent white, but even here the gains have usually been mixed, intermittent or nonsignificant. [St. John, 1975:119]

Research findings on the effects of school integration on educational aspirations, an important intervening variable in educational attainment models, are equally inconclusive. Some evidence suggests that biracial schooling lowers the disadvantaged students' educational aspirations. Plausibly, for some students integration with middle-class white students leads to a diminished sense of their own ability and self-worth and as a result lowered aspirations. Indeed, the "community control" concept of local schools as an approach to equality of urban education stemmed, in part, from the observation that school desegregation inevitably implies that poor and minority students can learn only by sitting next to well-to-do white students.

The heated controversy over the implications of recent findings clearly does not affect the moral case for integration. The other major value that has been sought in the school desegregation policy was to provide an environment that children can perceive as representing the American society, and in which they are socialized into the core ideals of the nation. The impact of school integration on the social system cannot be evaluated by test scores and academic achievement, no matter how conclusive the findings might be.

EXPECTATIONS VS. PERFORMANCE

Our urban educational problems have not been solved through the involvement of the federal government. Despite federal aid to education, the educational attainment of children attending schools in the suburbs exceeds that of the poor and minority students. This, however, does not imply that some programs were not successful and that others were total failures. Furthermore, as discussed in the previous section, the evaluations of many programs did not furnish conclusive results. Nevertheless, the high expectations associated with the federal role in urban education have not, generally speaking, been realized. In this section some major interpretations of the gap between expectations and performance are discussed.

THE METHODOLOGICAL CRITIQUE

Program evaluation is the newest social science discipline and in many ways it is in a state of development. Early evaluations tended not to appreciate at the outset the difficulty of alleviating the complex problems associated with educational attainment among poor and minority students. The underlying assumption in many educational evaluations has been that the program impact would be so great that it could easily be demonstrated, once the program goal(s) are agreed upon and amenable to measurement. The search for program impact is, however, much more complicated and involves a great number of considerations, including implementation, variability of the program variables, theoretical assumptions, research design, measurement, data collection and analysis, and the organization context of the research (Nachmias, 1979). When these considerations are taken into account, many of the findings reported so far are tentative at best, although their publicity has rendered a different impression.

Moreover, no-impact findings based on certain theoretical assumptions can be interpreted as positive under a different set of assumptions. Since most educational evaluations do not explicitly articulate the theoretical framework that guides research, their findings can be legitimately interpreted in different ways. For example, a behavioristic theoretical orientation postulates that all behavior is learned. This implies that educational inequities exist because specific academic or social skills necessary for educational achievement have not been acquired. These skills must be taught to the disadvantaged child using an appropriate instructional program. The success of the program is determined by tests that measure distribution of the target variable among program participants and on or more control groups. Obviously, no impact findings under such orientation is meaningless under the cognitive development orientation or the psychodynamic approach. The former emphasizes the process of cognitive growth more than the learning of a specific skill, and the latter stresses socioemotional goals to be essential for the development of the "whole" child, and postulates that the child "Knows what is best" for his personal growth.

THE NONINTERVENSIONIST VIEW

The nonintervensionist view sees the gap between expectations and performance as inevitable because disadvantaged children are believed not to have the ability to improve their cognitive skills and educational attainments. By implication, the government role in urban education is to be contracted to a minimum. This position is most closely related with the work of Arthur Jensen (1969). Essentially, the argument is that IQ is the most significant basis for learning skills, and that it is primarily inherited through genotype passed on from parents to children rather than being transmitted through environmental influences. The poor are impoverished for the same reasons that their children cannot learn; they have limited potential as manifested by their low IQ scores. Government programs, no matter how sound and well-implemented, simply cannot be expected either to enhance the cognitive skills and educational attainments of disadvantaged children or their social mobility.

In the nonintervensionist view, questions of ideology, theory, and evidence are inextricably interwinded. Basic ideological convictions about human nature and the human potential shape the interpretations of why some students succeed in schools and others fail. Empirically, however, this view is flawed on a major count. The assumption of constant intelligence is challenged by educational programs that have experienced sharp increases in IQ scores during the duration of the programs, as discussed earlier. The termination of the programs reverted IQ scores back to the lower levels. Certainly, this does not call for the termination of programs but their continuation for longer periods and beyond the preschool and the first classes of elementary schools.

THE INTERVENSIONIST VIEW

Several explanations for the gap between expectations and performance are offered by supporters of government intervention in urban education. First, it is suggested that while federal educational policies have only slightly improved education in urban schools and the educational attainment of disadvantaged children, the gap between expectations and performance should have been anticipated, given the serious and complex problems associated with educational and occupational attainment. Moreover the problems of urban education were not likely to be solved by state and local efforts. Without the committment of the federal government and utilization of federal funds, the problems would have been worsened.

Proponents of government intervention also maintain that time is a crucial condition, both for individual development and organizational change. Thus, crash educational programs are likely to fail; the federal government must authorize and support educational programs of long duration. Furthermore, it is argued that adequate resources have not been provided. A fundamental change in urban education requires the allocation of far more resources than were thought to be necessary when the programs were developed. The more effective programs were those in which funds were concentrated on a small enough number of children to make adequate support possible for whatever school and home activities were designed.

Finally, most programs designed to increase educational attainment have focused primarily on cognitive skill, and academic performance. However, the analysis of the status attainment process indicates that their role is far from simple and direct. Their effect tends to be mediated largely by social-psychological factors, which, in turn, also have an independent influence on the process of educational attainment. Very little attention has been given to these factors in most of the programs reviewed. The major implication of the status attainment research is that programs designed to manipulate the social-psychological environment of the disadvantaged student would have important effects on status attainment. More resources should have been directed at programs designed to influence the expectations of teachers or programs which attempt to acquaint parents with the academic potential of their child, and to make them aware of the importance of academic achievement to later educational and occupational opportunities.

PROSPECTS FOR THE 1980s

The economic and political conditions that induced the federal government to take a more active role in equalizing educational opportunities are rapidly changing. Most significantly, attempts to contract the role of government are increasing. These range from legislative and executive branch initiatives to direct referendum initiatives in states and municipalities. At the same time, the prospects for slower economic growth are being felt in the job market. The economic value of a college degree, measured by the comparative starting salaries of college graduates and nongraduates has somewhat declined (Glazer, 1975). The view that college graduation brings with it an assurance of multiple and renumerative jobs is increasingly being challenged. These developments are especially pronounced in the cities where the range and the magnitude of the nation's social problems are more intensely concentrated. But the cities have been losing their prominent place on the public's and policy-makers' agenda, making them more vulnerable to the changing economic and political conditions.

In the 1960s the belief that education is the answer to urban problems, including poverty, was widespread. The relationships between lower levels of education, higher rate of unemployment, and lower levels of income led to the belief that increasing the education and skill levels of the urban labor force would solve the cities' problems.

> The best long term bet (for combating poverty), we think, is simply education. Despite some unencouraging studies of compensatory programs, experience suggests education is a better engine of social advancement than any alternative. And partly because too much energy has been wasted arguing other things, the nation's cities have yet to evolve a truely comprehensive and coordinated school program for students. Such programs would require spending more money on slum schools than even on suburban ones. [Wall Street Journal, 1967]

With declining rates of economic growth and reduced employment opportunities for college graduates, quality of education rather than educational attainment per se would play a more significant role in occupational attainment and social mobility. This, in turn, would place urban schools in a less competitive position vis-à-vis suburban schools, unless governments at all levels were to increase their commitments to educational programs that either proved successful or resulted in inconclusive findings. If, as the urban poor extend their educational attainment, those who are better off extend theirs, the gap between the two will not be reduced, unless educational resources are further redistributed as well as increased.

The inconclusive state of many urban educational programs should not necessarily be interpreted as a failure to delineate the right problem. While in the Senate, Walter Mondale, introducing the Full Opportunity Social Accounting Act of 1967, reviewed a variety of government programs and concluded in the following way:

> Our intentions are good, but we lack the systematic and integrated approach to social problems. When they miss their mark, we may not even realize it. If they are damaging, it may be years before we know it. Our successes are difficult to document; they suffer the attacks of the ignorant while we, in our ignorance, have no way to defend them. [Mondale, 1967:34]

In developing its urban educational policy, the federal government focused on two important factors in educational attainment: enhancing the cognitive skills of the urban poor and increasing their years of schooling. These, however, are only two factors in the complex process of educational and occupational attainment. Programs aimed at enhancing cognitive skills and years of schooling, no matter how successful, can improve, but not solve the complex educational and occupational problems of the cities. The educational attainment model described earlier, is characterized by an individualistic orientation focusing predominantly on the impact of social origin and individual characteristics such as intelligence or educational attainment on occupation and income and on intergenerational mobility. There is only a limited concern for extraindividual or structural constraints in the opportunity structure, such as limited access to quality schooling and discrimination in the job market.

The belief that education will solve the problems of the urban poor, and that educational attainment can be changed by increasing levels of motivation, ability, and skills, is based primarily on a neo-classical, free market conception of occupational placement. "The competivite market assumed by neo-classical economic theories guarantees that the differential placement of individuals in the socio-economic order is a reflection of the individual characteristics brought into the market place by the worker" (Horan, 1978).

Even a comprehensive educational policy dictated solely by this perspective cannot be expected to solve the problems of the urban poor. The lack of availability of occupational opportunities in an era of slow economic growth compiled with racial discrimination, are two important factors that intervene between education and income. This calls for a change in the theoretical orientation to status attainment.

Models of status attainment are currently structured as if the occupational market place is wide open, with occupational level being a function of one's talents, ambitions, and credentials. This was not the case in the 1960s and it will be even less so in the 1980s. Thus, even if one could meet the educational needs of the urban poor, education would have only a limited impact on the future stratification positions of these youth. The know-how which is required in addition to college graduation for many positions, the racial discrimination which occurs in the recruitment and promotion of employees and the uneven rewards of various levels of educational attainment intervene between education and social mobility. There is a need for more emphasis on the structural limitations and the importance of societal factors in the selection and assignment of individuals to status positions. Externally imposed limitations on attainment, such as racial discrimination practiced by social agencies and institutions, will have to be included in models of status attainment with attempts to study its base, timing, or resources (Kerckhoff, 1976). Moreover, this calls for a more comprehensive and integrated federal educational and urban policy. Government commitment to existing educational programs will not be adequate in an era of limited opportunities. While programs aimed at individual change will require continued support to sustain the limited gains that were made, a concentrated effort is needed to assure the removal of structural barriers that limit access to occupational opportunites at a given level of educational attainment.

NOTE

1. Obviously there are other measures of program success not directly or indirectly affecting earnings. Improved nutrition and self-esteem are but two examples.

REFERENCES

BLAU, P. M., and O. D. DUNCAN (1967) The American Occupational Structure. New York: John Wiley.
CARTER, T. M., J. S. PICOUS, E. W. CURRY, and G. S. TRACY (1972) "Black-white differences in the development of educational and occupational aspiration levels." Unpublished manuscript, University of Minnesota.
COLEMAN, J., E. Q. CAMPBELL, C. J. HOBSON, J. McPARTLAND, A. M. MOOD, F. D. WEINFELD, and R. L. YORK (1966) Equality of Educational Opportunity. Washington, DC: U.S. Government Printing Office.
GLAZER, N. (1975) "Who wants higher education even when it's free." Public Interest (spring): 130-135.
GOLDBERG, M. L. (1963) "Factors affecting educational attainment in depressed urban areas," in A. H. Passow (ed.) Education in Depressed Areas. New York: Teachers College Press.
GRAY, S. (1974) U.S. Department of Health, Education and Welfare, Office of Human Development, Longitudinal Evaluations: A Report on Longitudinal Evaluations of Preschool Programs, ed. Sally Ryan, Vol. 1, (OHD) 72-24. Washington, DC: Department of Health, Education and Welfare.
HANSEN, L. W. and R. J. LAMPAM (1974) "Basic opportunity grants for higher education." Challenge (November-December): 46-51.
HORAN, P. M. (1978) "Is status attainment research atheoretical?" Amer. Soc. Rev. 43 (August): 534-541.
HOUT, M. and W. R. MORGAN (1975) "Race and sex variations in the causes of expected attainments of high school seniors." Amer. J. of Sociology 81 (September): 356-363.
HURN, C. J. (1978) The Limits and Possibilities of Schooling. Boston: Allyn and Bacon.
JENSEN, A. R. (1969) "How much can we boost IQ and scholastic achievement." Harvard Educ. Rev. 39: 1-123.
KERCKHOFF, A. C. (1976) "The status attainment process: socialization or allocation?" Social Forces 55: 368-381.
––– and R. T. CAMPBELL (1977) "Black-white differences in the educational attainment process." Sociology of Education 50 (January): 15-27.
LEVIN, H. M. (1978) "A decade of policy developments in improving education and training for low income populations." Evaluation Studies Review Annual 3: 521-570.
LIPSET, S. M. and H. L. ZETTERBERG (1956) "A theory of social mobility." Transactions of the Third World Congress of Sociology 2: 155-177.
MONDALE, W. (1967) U.S. Senate, 90th Congress, Hearings before the Subcommittee on Government Research of the Committee on Government Operations, Full Opportunity and Social Act Seminars. Washington, DC: U.S. Government Printing Office.
NACHMIAS, D. (1979) Public Policy Evaluation. New York: St. Martin's.
PICARIELLO, H. (1969) "Evaluation of Title I." U.S. Office of Education, Office of Program, Planning and Evaluation.

PORTES, A. and K. WILSON (1976) "Black-white differences in educational attainment." Amer. Soc. Rev. 41 (June): 414-431.
RIBICH, T. and J. MURPHY (1975) "The economic returns to increased educational spending." J. of Human Resources 10 (winter): 56-77.
SEWELL, W. H. (1971) "Inequality of opportunity for higher education." Amer. Soc. Rev. 36 (October): 793-809.
––– and R. M. HAUSER (1975) Education, Occupation and Earnings. New York: Academic Press.
SORKIN, A. L. (1974) Education, Unemployment and Economic Growth. Lexington, MA: D.C. Heath.
SPILERMAN, S. (1971) "Raising academic motivation in lower class adolescents: a convergence of two research traditions." Sociology of Education 44 (winter): 103-118.
St. JOHN, N. (1975) School Desegregation: Outcomes for Children. New York: John Wiley.
STEBBINS, L. B., R. G. St. PIERRE, E. C. PROPER, R. B. ANDERSON, and T. R. CERVA (1978) "An evaluation of follow through." Evaluation Studies Review Annual 3: 571-610.
Wall Street Journal (1967) November 15.

13

Health Care Policy: Disillusion and Confusion

ANN LENNARSON GREER

INTRODUCTION

☐ BY 1977, THE HEALTH SECTOR of the national economy, which consumed only 3.8% of the GNP in 1930, had become the third largest industry in the United States. In that year, health costs reached $163 billion, a threefold increase over 1970. This total constituted 8.8% of the nation's GNP, up from 8.3% in 1975. Experts predict that the proportion of the GNP consumed by health costs may rise as high as 12% by the year 2000 (Public Services Laboratory, 1978).

As we move into the 1980s, the focus of federal health policy is cost containment, seen as both holding down the price of services and restraining the expansion of the health system. Ironically, much of the current cost containment effort is directed to halting the momentum or mitigating the effects of federal programs of the 1950s and 1960s, the primary purpose of which was to expand the medical care system and increase its utilization. Expansion and extension were thought to be essential to assuring the access of all Americans to the life-saving techniques of medical science. It is not difficult for any of us to remember when it was in poor taste at best and blasphemous at worst to even speak of the price of saving lives.

AUTHOR'S NOTE: *I wish to acknowledge my deep gratitude to Daniel I. Zwick for his comments on a draft of this paper.*

Viewed as policy, current containment efforts have much less coherence than did the expansion programs. We seem to be throwing up hastily constructed dikes which we hope will hold back a flood of costly health services. Health policy discussions portray the medical system as glutted with hospital beds, burdened by excess technology, and paying too many doctors to provide too much medicine. How is it that we now appear to be oversupplied with the very resources which we set out to create so short a time ago? The path we have traveled since we optimistically attacked illness, along with poverty, is the subject of this paper.

THE PROGRAMS OF THE 1960s

In order to make more and better care available to more and more Americans, a variety of programs were begun in the 1960s. The catchwords of the day were quality, availability, and accessibility. Quality usually meant that services should be characterized by the most up-to-date medical science, although a number of organizational innovations were also initiated in this period. Availability was a problem of the total quantity of health resources in existence. Accessibility was perceived as a matter of "obtainability" i.e., the geographic and cultural "closeness" or "openness" of services and assuring the individual's ability to pay for them.

The attack on *quality* of care involved research and education. Billions of dollars were invested through the National Institutes of Health to support research in the medical sciences. To achieve more rapid and widespread use of the latest in scientific knowledge and technology, Regional Medical Programs (RMPs) were established across the country. Closely affiliated with or run by medical schools, these programs undertook extensive outreach education and related activities to update the knowledge of local physicians and to acquaint them with the latest technologies. Through these programs, it was hoped that physicians away from the centers of science and development would bring local practice closer to the state of the art achieved at such centers.

Especially important among the programs designed to increase the quantity and therefore the *availability* of health resources were:

(1) capitation, construction, and other programs which provided funds to institutions training health professionals, especially

physicians, to help cover the costs of increasing their class sizes and the number of their graduates;

(2) construction grants under the Hill-Burton Hospital Construction Act to permit the construction of new hospitals, especially in rural areas, and the renovation and enlargement of existing hospitals.

It was hoped that by addressing the quantity issue, these programs would also address problems of *accessibility*. By increasing the number of medical school graduates, policy-makers hoped to foster a "spillover" of physicians from the more attractive practice areas into the less preferred rural and poverty areas. Similarly, it was hoped that the construction and renovation of hospitals in underserved areas would influence the locational choices of medical personnel since some of the disadvantage of underserved especially rural areas in attracting them was thought to be due to lack of modern hospital facilities.

Other programs addressed problems of accessibility resulting from maldistribution of medical resources. The National Health Service Corps Scholarship Program provided scholarships to medical and other students who would agree to "pay back" their scholarships through limited-time service in underserved areas subsequent to graduation. The rarely fulfilled hope was that these young physicians would put down roots in the underserved areas and stay beyond the pay-back period. The desire to see corps physicians remain in underserved areas led, in some schools, to preferential treatment for scholarship applicants who were themselves from underserved areas in the hope that personal attachments might lead a higher percentage to make the return permanent.

The most direct attack on the problem of maldistribution of resources was the Neighborhood Health Centers program initiated within the Office of Economic Opportunity. This program supported the creation of clinics in underserved communities. Neighborhood Health Centers (NHCs) were operated largely through federal grants made to nonprofit corporations, the governing boards of which were required to contain representatives of the poor and socially disadvantaged members of the communities served by the NHCs. These boards represented an attempt to assure that services were not only geographically accessible, but also that they were attuned to the special needs and cultural preferences of the poor, the aged, and the ethnically handicapped. NHCs also incorporated ideas thought to address problems of improper utilization, such as team care, wherein attempts were made to assure that care was comprehensive and continuous, rather than partial and sporadic. Neighborhood Health Centers were frequently housed in buildings constructed with federal aid. The National Health Service Corps became an important source of their physician staff.

Most visible to the public of the programs attacking health needs were those which provided for payment of services through Medicare and Medicaid. To many, these programs appeared to be predecessor programs to a government-administered National Health Insurance. We would begin with the most needy, the old and the poor, and later, with the resources in place, extend coverage to the citizenry as a whole. Certain deductibles and limitations were written into these laws, with Medicaid coverage varying widely by state, but for the wide range of hospital services covered, reimbursement was for "reasonable" costs. Meanwhile, coverage under Blue Cross and private health insurance expanded dramatically among workers and the more affluent.

While widespread medical insurance is relatively new, there has been little market behavior in medical services for many years (Stevens, 1971). Market competition was reduced to near zero by the 1910 Flexner Report which drove "quack" medical schools out of the market; state support of licensure by the remaining medical schools; increasing faith in the "scientific medicine" practiced by their products; and guild prohibitions against advertising. In brief, there was little information available for the consumer on comparative costs, and, especially, quality of service. The American Medical Association and state licensing bureaus were of no help. Their record in expulsion of incompetents was puerile. Blue Cross, a creation of hospitals interested in assuring payment of their charges, took little interest in what the charges were, nor did it or other insurers have the capability to determine value (Law, 1974). With no information on price for given quantities of service at given levels of quality, neither the customer nor the customer's insurer is able to "shop."

In view of this history, it was not surprising that cost reimbursement was adopted by Medicare/Medicaid, Blue Cross/Blue Shield and private insurers. Assurance of reimbursement for costs incurred in the provisions of medical care without respect to demand for or utilization of services stimulated an expansion unchecked by the market. Physicians, who to a great extent define the need for, as well as the price of, their services, did not find it necessary to "spill over" into underserved areas.

Hospitals which received reimbursement for their costs (i.e., their expenditures) could develop new services without minimum risk about utilization. For the public and private health insurers, each expansion of services increased costs dramatically.

Why did we fail to foresee and refuse open-ended costs as we increased the number of doctors and hospitals? That neither the market nor a substitute in regulation was seen as a need during the expansive period may be attributed, I believe, to certain widely shared assumptions. It is the questioning of those assumptions and not simply unexpectedly high costs which explains our current willingness to shift attention from the extension of care to the containment of its costs.

CHANGING ASSUMPTIONS

In the 1950s and 1960s when the expansive programs were introduced, the American public and its policy-makers looked backward to three decades of genuinely inspiring medical advances in prevention, diagnosis, pharmacology, and surgery. They looked ahead to continued national prosperity. We prided ourselves, perhaps above all else, on our science, our technology, and our American "know-how." It was these features of our national character which we felt had made us world leaders in agriculture, industry, and medicine. In the postwar years, the United States differentiated itself from other Western democracies, which were establishing national health systems to deliver care, by investing its big dollars in biomedical research and private health insurance. Our belief in science and technology was belief also in the competence of credentialed professionals. Year after year, physicians and scientists were rivaled only by Supreme Court justices for the top spots in national polls of occupational prestige.

The latter sixties and the whole of the seventies were characterized by a battering of our confidence and optimism. The brightest and best credentialed failed us in Vietnam. Top-ranked economists proved hard pressed to explain, let alone control, inflation or protect our currency. The very basis of our technology-dependent society was called into question by the energy crisis. Throughout the many sectors and institutions of our society, assumptions of expansion and progress were challenged by nay-sayers, who argued that we might have to live within existing energy supplies, without real growth in our economy or personal incomes, and with whatever quality air we could manage to protect.

In medicine, huge investments in research were failing to produce miracles comparable to the cures and protective vaccines which had inspired earlier enthusiasm. A cure for cancer did not develop easily. Chronic illnesses increased. The best that biomedical researchers seemed now able to produce were life-sustaining machines. Lewis Thomas (1974) called these expensive but scientifically unsatisfying contributions "half-way technologies" contrasting them to truly "high technology," that which prevents or cures disease. Yet life-sustaining mechanisms, such as respirators, kidney machines, and transplanted aortal valves, came into wide use. It was largely these alternatives to cure, rather than cure, which spread throughout the country, encouraged and paid for by federal programs. At first, it seemed these were devices to keep us breathing while we awaited cure. Imperceptibly, cure lost its prominence in our discussions and the machines and procedures which perpetuated existence came to dominate attention.

Distressingly, unlike the technologies which had increased productivity and expanded national wealth, these medical technologies began to look like they might bankrupt us. Extremely expensive, they require continual or recurrent application. They fail to free recipients of disease. Adding insult to injury, medical researchers and practitioners differ in their assessment of the appropriate use of these costly treatments. Differing medical judgments have themselves become a focus of attention. Health insurers have begun to encourage and pay for second opinions when elective surgery is recommended. Practicing physicians are often cynical about claims made for new procedures and machines (A. L. Greer and Zakhar, 1977, 1978).

Increasing loss of confidence in science, in technology, and the ability of the medical professional to interpret and apply them is signaled in other ways. The idea that disease and dysfunction can be "iatrogenic" (a result of medical treatment) has become commonplace. Physicians, federal agencies, and consumer advocates advise continuously that drugs have dangerous side-effects, that the risk associated with surgery may outweigh possible benefits, and that germs are picked up in hospitals. How quaint it now seems to find Webster's 1965 dictionary describing "iatrogenic" as a term "used chiefly of imagined ailments."

NEW APPROACHES

This loss of faith has been accompanied by a searching for new paradigms within which to think about illness and its treatment (Dubos, 1959, Fuchs, 1974; Carlson, 1975; Knowles, 1977). Researchers and policy-makers question whether the diseases which now dominate our attention are susceptible to dramatic intervention. Increasingly, prevalent diseases are attributed to heredity, age, and unhealthy living, not to germs or bacteria. An increasingly larger portion of today's ill, it is noted, suffer from chronic diseases. Patients entering acute care hospitals frequently depart not for their private residences, but for nursing homes.

"Epidemiology" is an approach receiving greatly increased attention. Its practitioners examine distributions of disease in the population—partly to assist planners seeking formulae for locating traditional health services, but more importantly as a means to identify sources of dysfunction in lifestyle, genetic inheritance, or environmental risks.

At a level of immediate action are advocates who feel we know enough about the sources of today's maladies to propose alternatives to high-priced hospital care (Belloc and Breslow, 1972; Belloc, 1973; LaLonde, 1975). Watchwords are "wellness" and "self-care." Government should concern itself with behavioral and environmental factors contributing to health or ill health. People who wish to be healthy should live healthy lives—avoid tobacco and excessive liquor, eat nutritious diets, and exercise. Traditional health services should be incorporated into a solid health regimen—not substituted for one. The individual should assume responsibility for maintaining personal health and not expect technology (and public money) to undo self-inflicted damage.

A related approach is that which proposes "disease management" as the treatment of choice for individuals suffering from chronic disease. For example, physicians such as Ehrlich (1973) are now suggesting that a person suffering from rheumatoid arthritis should be taught principles of movement which will minimize joint damage, and thereby slow the progression of the disability. Powerful drugs and surgical replacement of joints should be viewed as treatments of last resort only. "Disease management" as a treatment modality combines awareness of risks associated with treatment, especially surgery, with ideas of "self-care" and "wellness."

An approach which was concretized into a principal recommendation of the President's Commission on Mental Health (1978) disputes the value of the monopoly role of professionals in providing care. The commission recommended greater reliance on "community support systems." The latter refers to the networks of family, friendship, church, and community which have traditionally assisted individuals in time of need and cared for the chronically ill. Although documentation of the current availability of such networks has lagged behind enthusiasm for their resurgence, proponents cite the performance record of Alcoholics Anonymous and the success of some patients, such as those suffering kidney disease, in treating themselves at home. Care provided by family and friends is held to be more humane than that provided by professionals; care at home is to be preferred over care in institutions.

Finally, receiving much attention are those (see Ross, 1969) who declare explicitly that death is the expected and perhaps desired outcome of many illnesses. The spread of hospices for the terminally ill, along with training for social workers who would assist persons and families facing death are upshots. Legal battles concerning the right to die at home among family and friends and disconnected from life-prolonging machines evidence a dramatic departure from our past faith in technology.

As an approach to illness, increased emphasis on "epidemiology" constitutes a departure from our emphasis on biomedical research. Those advocating "community support systems" and death with dignity question the extent to which professionals hold special keys. "Wellness," "self-care," and "disease management" are strategies, similar, I believe, to those now being proposed in other sectors of activity. With respect to energy resources, economic growth, personal finances, and health, Americans are increasingly viewing their situation in terms of scarcity, stagnation, resource management, and coping. Whether this mood reflects a more or less realistic assessment of our science or our future, its diffusion marks a retreat from our once higher expectations.

THE PROGRAMS OF THE 1970s

Cynics are quick to point out that low-cost, even no-cost health care alternatives have become topical synchronous to a widely espoused desire to hold down government costs. Discussion of alternative approaches to health care have not led to the funding of new programs. The Presidents Commission on Mental Health with its reliance on community support systems was able to confine itself to recommendations totaling only an additional $500 million over five years for all federally supported mental health services. Discussion of federally financed National Health Insurance has declined. If there is to be government action to insure health care average, it now appears it will be through requirements placed upon employers. Policy-makers remain absorbed in the problem of containing the costs of the existing medical system. In this state of ideological confusion, the question has become: To what extent can we continue to pour money into a health system with huge costs and disputed effectiveness?

Thus, the programs of the 1970s are programs of containment and stabilization. Those now operational were created by legislation passed prior to 1975 and contain language of the sixties as well as the seventies. The objectives of quality, availability, and accessibility, however, have received short shrift compared to cost containment. The organizations charged with their achievement have faltered or been pressed with demands to give primary attention to cost containment. A few major examples may be cited.

Created by legislation passed in 1973, health maintenance organizations were to enroll families in prepaid health plans providing preventive as well as curative care. Many benefits were anticipated. For example, families enrolled in prepaid plans were expected to obtain the comprehensive and continuous care which would lead to good health and prevention of serious illness and avoidable hospitalization. Prepayment would remove incentives for physicians to "overdoctor" especially to perform questionable surgery. Because both of these characteristics contained the potential of reducing the frequency of costly hospitalization and expensive procedures, the hope was that care obtained from health maintenance organizations would not only be of higher quality but would, in the long run, be cheaper. The potential cost savings of HMO care were always a source of controversy, however, with some claiming that people would obtain more care and drive costs up. Determination to learn the long-run cost-benefit of care which may be initially more expensive lost much of its attractiveness amid the focus on short-run costs and declining interest in quality of care experiments. HMOs have, in fact, captured little of the medical care business. Perhaps most important in their slow growth is the fact that they impose restrictions on consumer choice of physician and hospital. Although corporations offering health insurance to their employees must offer an HMO alternative if one is available, families have tended to opt for unrestricted coverage, which is almost universally offered by insuring employers.

Two umbrella programs which were created to promote proper use of health services are professional standards review organizations (PSROs) and health systems agencies (HSAs). By law, these programs work through federally funded local organizations designated to perform their respective functions.

On the assumption that only physicians can judge or regulate the provision of physician-provided care, the PSRO legislation passed in 1972 required that organizations of local physicians compose themselves to monitor the quality and cost appropriateness of the physician care provided in hospitals (S. Greer, 1978). PSROs were to create a mechanism for comparing an individual physician's behavior with local standards. In practice, this came to meaning monitoring the sequence of actions taken by a physician once a diagnosis was entered for (1) medical soundness, and (2) unnecessary hospitalization. In the spirit of the 1970s, the quality concerns of the PSRO act have been deemphasized. PSROs have emerged largely as arbiters of appropriate length of each hospital stay.

Over 200 federally funded health systems agencies now blanket the country, charged with the responsibility for health system planning in their areas—an activity which, in principle, includes both sponsoring and restraining activities. Restraining activities have been increasingly emphasized. The Health Planning and Resources Development Act of 1974, which created the HSAs, included provision for area health development funds to be spent by the HSAs in encouraging innovative programs, but funds have never been allocated. HSAs have proceeded to develop multifaceted plans for their areas, but health plan committees have suffered from widespread lack of community interest (unless plans include provisions for abortion). By contrast, committees charged with the review of proposals for hospital construction, expansion, renovation, purchase of machinery, or introduction of new services have been hotbeds of activity and controversy.

The health systems agencies are without regulatory powers themselves, but variously through law, regulation, and courtesy have come to figure in the deliberations of the increasing number of organizations which, through state power or control of insurance monies, must approve expansion of health services or rates for their reimbursements.

The reviews of health systems agencies, undertaken to advise state and federal agencies considering funding programs in the local area, often receive considerable media publicity. Negative recommendations may make approval at higher levels appear contrary to the public purpose. As individual states go beyond federally mandated restraints with experiments in such things as cost "caps" (as in oil wells), or "decertification" of existing services (i.e., taking away the approval which allows or blesses reimbursement by public and private payers), HSA reviews again play a role. Insofar as Blue Cross and Blue Shield increasingly are being impelled to focus on containing costs, they frequently look to HSAs for guidance or rationalization for resistance to increasing costs.

Some HSAs have become very controversial. Some of the bases of controversy were implicit in their structure. Governing bodies are composed of a mixture of local citizens including "providers" of care (physicians, hospital administrators, etc.) who presumably bring to the deliberations both expertise and self-interest, and "consumers" of health care, defined as citizens of the area who are not providers of care, and who are presumably not self-interested. Local, accessible and often highly visible, the deliberations and decisions of these boards and their feeder committees are forums of much community interest. While their recommendations are advisory only, they usually bring forth the first verdict on a proposal, and thus are targets of attempted suasion by those who would expand and those who would contain the particular areas's health institutions. Insofar as the battle between governmental agencies empowered to restrain and health care organizations and associations committed to expansion has proven itself to be a real one, i.e., a battle between powerful opponents, the HSA is a political hotbed, especially so since it tends to bring machinations of health politics into the view of a larger public.

THE PROGRAMS IN RETROSPECT

Not all of the programs of the 1960s were such howling successes as those which set out to increase the number of hospital beds and physicians and the amount and extensiveness of technology. We are plagued today hardly at all by the negative consequences of being too successful in redistributing health personnel or reorganizing their manner of work. President Nixon's deemphasis of the Neighborhood Health Centers Program occasioned nothing like the controversy and power politics which beset efforts to restrain the further growth of the health industry. Health maintenance organizations have languished. Programs which might identify new approaches to community care await even experimental funding.

The problems in health care organization which were to be addressed by neighborhood health centers and health maintenance organizations are with us today as they were fifteen years ago. These programs which depended for their success upon change in the behavior of professionals never saw the rapid implementation and growth of the programs embraced by medical schools, hospitals, and communities eager for expansion opportunities. The problems they would have attacked, such as comprehensiveness and continuity of care, persist. What has changed is our clear conviction that the nation's foremost health need is to get winning combinations of traditional health care providers to the scenes of our troubles.

Traditional reformers find themselves confronted by the same confusion in assumptions about health as confront the nonreformers. Their proposals suffer not only (although possibly primarily) from the fact that the traditional system's capacity to consume available resources appears cavernous and unstoppable. It is also true that choice in approach brings with it easy immobilization and a rationalization for immobility.

The continuing health needs of American minorities provide illustration. It is undisputed that the health status of certain identifiable groups, especially poor blacks, lags behind that of the population as a whole. Certain statistics for the black population, such as infant mortality, constitute a national embarrassment. What should be done? Explanations of these statistics are now numerous. Many continue to feel that the health status of the minority poor remains one of redistributing and reorganizing traditional health services. For example, in 1978 the prestigious Robert Wood Johnson Foundation launched a five-city demonstration project wherein selected cities received grants to establish primary care clinics in inner-city neighborhoods. Their primary objective is to provide continuous comprehensive *medical* care to inner-city residents. Similarly, the National Health Service Corps Scholarship Program with its ability to *direct* health personnel into needy areas is currently preferred by Congress to medical school capitation, since, unlike capitation, it constitutes a vehicle for continuing the attack on the problem of maldistribution of personnel.

Is this the answer? Eli Ginsberg (1978) has commented on reports that the poor now pay ambulatory visits to physicians more frequently on the average than do middle- and high-income families, and are also admitted to hospitals more frequently. The meaning of this is unclear, since in view of their generally poorer health, they may still be relatively deprived—and needy. He notes, however, that they may also be "seeking and receiving more medical attention (frequently of indifferent quality) than the optimum."

The high rates of infant mortality among black mothers focus the confusion. Rates of infant mortality are high where health services are scarce—in, for example, inner-city minority neighborhoods and the rural Southeast. Other areas lacking easy access to medical care, however, such as the upper Midwest, show remarkably low rates of infant mortality, and allow a questioning of the role of medical services, or at least the need for their geographic accessibility. To explain the discrepancy, epidemiologists suggest examination of differences in genetic inheritance, nutritional habits, lifestyle, and such things as the age at which mothers in particular populations give birth. They cannot rule out, however, the possibility that the differences are attributable to medical care received. It is possible that prospective mothers in the upper Midwest overcome obstacles imposed by distance to obtain high-quality medical care. We lack the knowledge to sort out explanations, put medical services into perspective, and devise a program of action.

Looking back on those programs which accomplished their objectives and those which did not, it is ironic that it was the establishment-supporting programs which most succeeded, i.e., those which expanded and allowed firm funding for the growing health industry and its close relations, the drug industry, the construction industry, the banking industry, the manufacturers of medical equipment, the universities, and so forth. In retrospect, however, it seems that there was really no challenge to the importance of these groups in the politics of the era. The "power to the people" aspects of the 1960s programs were appeals for greater access to the services of these highly valued professionals and access to their jobs. Professionals were charged with insensitivity (and were thus asked to work in clinics for poor people's boards which would ensure their greater responsiveness) but not with ignorance. A greater role was envisioned for paraprofessionals, but guided and supervised by those with established credentials. Many more minority people obtained access to middle-class jobs, including the health professions but their numbers remain low.

It is the state of our knowledge which troubles us today, along with the loss of our sense of confidence that our problems are imminently solvable. An enlarged coalition of powerful health industry interests spurred onward, although certainly not created, by the programs of the 1960s, lobbies successfully for continuation of the programs which have made possible their growth and prosperity within a context of great security. Taxpayers and their representatives struggle against ever escalating costs but lack the same sense of purpose. Lacking a sense of the desirable quantity of medical services, a sense of scientific possibility, or impetus for taking alternatives seriously, there is momentum only for stabilization.

It is noteworthy that confusion puts many more providers of physical health care in a situation similar to that of mental health care providers who have lived with plural approaches a good deal longer. In 1970, Elliot Freidson noted in his book *Professional Dominance* that the science upon which psychiatry rested was insufficiently respected to endow its practitioners with the dominant role enjoyed by other physicians in the provision of care. The superior ability of psychiatrists to diagnose and treat mental and emotional problems has been successfully challenged by psychologists, marriage counselors, child welfare professionals, social workers, vocational rehabilitation personnel and others. As members of the elite holding medical degrees, psychiatrists have shared in the favored status enjoyed by M.D.s in third-party reimbursement, but their challengers have been accorded the right to offer services in a wide variety of settings.

The paradigmatic confusion creeping into physical health probably shares an important common origin with that which has characterized mental health: chronicity of disease. As noted earlier, medical science is at its least persuasive when invoked to treat the chronically ill. In mental health where chronic patients have occupied attention for a much longer time, the inability of medicine to produce dramatic results has probably fostered two other developments—both of which are now appearing in physical health also. The first of these is concern with prevention as an alternative to treatment. The second is a general confusion between health and social services.

For the chronically ill, medical care is but one component in a complex configuration of needs which include income, housing, housekeeping services, recreation, transporation, and so forth. That means are required to address these many needs of the chronically ill has not escaped the attention of the managers of hospitals looking for expanded roles as "life care" centers. The dispute in mental health areas as to whether life care needs should be defined as reimbursable health care or treated as social services can be expected increasingly to arise in "physical" health policy discussion. It is more likely that we will limit to some satisfactory level the expansion of acute care hospital beds and life-rescuing technologies than that we will easily solve the problem of what to do (at what cost) with not only those persisting on life-prolonging machines, but with the chronically infirm—a group whose size grows as our population ages.

WHO SHALL DECIDE: THE PROBLEM OF GOVERNANCE

Amid this general confusion, the federal government has retreated from delineating policies explicitly in favor of local determination. In recent years, local nonprofit citizen boards have been the most common designees. The Health Systems Agency boards composed of specified combinations of "providers" and "consumers" are typical. In designating these governing agents, the federal government has limited itself to the sketchiest of requirements—typically that boards contain 51% "consumers" or "citizens" and that they represent demographically the people who live in the region or catchment area. Language which has accompanied recent laws or amendments perpetuating such governing boards tends to generalities concerning "partnerships" between levels of government, providers and consumers, and the public and private sectors.

Citizen boards can be traced to numerous possible origins: traditional voluntary boards, appointed commissions, and the grass-roots people's boards of the 1960s. Whereas the reason for having one or another of these boards may at one time have been clear (ability to raise money for a hospital, presumptive freedom from corrupt political influence, or representativeness of recipients of services), it certainly no longer is.

The future of our current decentralization will be interesting—as local designees are faced with thorny issues. Confusions may be eliminated if higher levels of government impose reimbursement policies which have the effect of eliminating local choice. If not, it would seem that several important intellectual dilemmas related to local determination require resolution. The first concerns the legitimacy of the federally designated governing bodies now making decisions. With increasing belligerence, state and local governments have challenged nonprofit citizen boards as unaccountable and illegitimate. If they are intended to be political stand-ins for representatives in Washington who feel themselves too remote from local situations to make locally acceptable decisions, composition for legitimacy is of utmost importance. Electorally accountable local officials might be selected as most legitimate, as might blue-ribbon citizen commissions or demographically representative boards, depending upon the theory of legitimacy favored. The choice of politically accountable elected bodies like city councils or county boards has the advantage of "building on existing resources"—an idea popular in these days of resource constraint. Drawbacks from the viewpoint of some Washington analysts continue to be those put forth in the 1960s. Local government is claimed to be frequently incompetent and is accused of not being truly representative.

Another way to build on existing resources, if "blue-ribbon" legitimacy is acceptable, is to recruit influential leaders from the private sector, and hope they can attract private resources as well as funding from state and local government. That the federal government will strive in the creation of future boards for representatives of clients where this means poor people seems unlikely, since it also implies a willingness to be a permanent source of funds. A logical if little discussed possibility if medical benefits continue to come from employer/employee benefits is the German model where the financial "contributors" to health plans, labor and management, negotiate rates and services with providers (Lansberger, 1978). Specific links to health maintenance organizations might be a vehicle sought by industries in the future as they seek greater control over the size of the bill they will be paying. (Kaiser Permanente has already shown the way.)

The accountability least related to doctrines from political theory but acutely felt by many governing board members is accountability to the health status of the population. Such "accountability" however requires knowledge of achievable ends along with the costs of attaining them. No governing body has this information. We have experienced the trap of specifying means without measurement of ends: the programs of the sixties very successfully produced more doctors, more hospital beds, and more technology while failing to produce the improvements in health which it was thought would accompany these achievements. We seem now to be searching for a political formula, like the U.S. Constitution, which will allow us to persist without knowledge or money. Our previous posture of assuming professionals would create the product we sought at a price we would be willing to pay now seems naive. Our current view of the power of science and professional expertise is no doubt more balanced. Yet, we are now in danger of accepting as our major task the dividing up of units of geographically accessible health care resources or the right to construct or close buildings.

The need for knowledge at all levels has never been greater. Our political bodies, at whatever level they exist, have a duty to make meaningful decisions, not simply to divide up a shrinking pie. To choose between spending scarce dollars on preventive care to improve the likelihood our children will be healthy adults, or restoring our currently productive population to good health, or making the aged more comfortable are worthy political choices. Without vastly greater knowledge of what present techniques can produce and at what cost, such choices cannot be made.

REFERENCES

BELLOC, N. (1973) "Relationship of health practices and mortality." Preventive Medicine 2: 67-81.
BELLOC, N. and L. BRESLOW (1972) "Relationship of physician health status and health practices." Preventive Medicine 1: 409-421.
CARLSON, R. J. (1975) The End of Medicine. New York: John Wiley.
DUBOS, R. J. (1959) The Mirage of Health. New York: Harper and Row.
EHRLICH, G. [ed.] (1973) Total Management of the Arthritic Patient. Philadelphia: J. B. Lippincott.
FREIDSON, E. (1970) Professional Dominance: The Social Structure of Medical Care. Chicago: Aldine.
FUCHS, V. R. (1974) Who Shall Live? Health Economics and Social Change. New York: Basic Books.
GINSBERG, E. (1978) "How much will U.S. medicine change in the decade ahead?" Annals of Internal Medicine 89 (October): 557-564.
GREER, A. L. [with the assistance of A. Zakhar] (1977) Hospital Adoption of Medical Technology: A Preliminary Investigation Into Hospital Decision Making. Urban Research Center Publication, University of Wisconsin–Milwaukee.
――― and A. ZAKHAR (1978) "The role of physicians in the introduction of medical technologies into hospitals." Presented at the 73rd Annual Meetings

of the American Sociological Association, San Francisco, September 4-8.
GREER, S. (1978) "Professional self regulation in the public interest: the intellectual politics of PSRO," in S. Greer, R. Hedlund and J. Gibson (eds.) Accountability in Urban Society: Public Agencies Under Fire. Urban Affairs Annual Review 15. Beverly Hills: Sage Publications.
KNOWLES, J. H. (1977) Doing Better and Feeling Worse; Health in the U.S. New York: W. W. Norton.
LaLONDE, M. (1975) A New Perspective on the Health of Canadians. Ottawa, Canada: The Government of Canada.
LANSBERGER, H. A. (1978) The Trend Toward "Citizen participation" in the Welfare State: Countervailing Power to the Professions?" Presented at the Ninth World Congress of Sociology (Workshop 10), Uppsala, Sweden, August 14-19.
LAW, S. (1974) Blue Cross: What Went Wrong? New Haven, CT: Yale University Press.
Public Service Laboratory, Georgetown University (1978) "Costs of disease and illness in the United States in the year 2000." A Special Supplement to Public Health Reports 93, 5: 494-588.
ROSS, E. K. (1969) On Death and Dying. New York: Macmillan.
STEVENS, R. (1971) American Medicine and the Public Interest. New Haven, CT: Yale University Press.
The President's Commission on Mental Health (1978) Report to the President. Washington, DC: U.S. Government Printing Office.
THOMAS, L. (1974) The Lives of a Cell: Notes of a Biology Watcher. New York: Viking.

14

City Personnel: Issues for the 1980s

DAVID H. ROSENBLOOM

☐ THE CONTINUING "URBAN CRISIS" and the increasing likelihood of widespread retrenchment by local governments raise very fundamental issues for urban public personnel administration. Indeed, it would not be difficult to conclude that personnel are at the center of the contemporary urban scene. Crime, riots, and decay are frequently in the limelight of discussions of urban America, but it is personnel costs that account for most of the expenditures to be found in the typical city's operating budget. In fact, it is not unusual for such costs to represent 60% to 80% of such a budget (Stanley, 1977: 31). In 1975 local agencies employed the equivalent of some 7,369,000 full-time personnel, an increase of 75% from 1960 and 18% since 1970 (Thompson, 1978: 83). By contrast, in 1976 employee costs accounted for 11% of the federal budget (Shafritz et al., 1978: 29), and their number actually declined slightly during the first half of the 1970s (Department of Commerce, 1976: 395, table 619). Moreover, in the view of some analysts, it is the nature of urban personnel administration—especially collective bargaining—that has brought cities to the brink of bankruptcy (Raskin, 1972). Thus, although 1978 witnessed the enactment of far-reaching reforms of federal personnel administration, it is at the local level that a major rethinking of public personnel administration would appear to be most critical. As retrenchment becomes more common, such an effort is inevitable and will have to come to grips with the problem of competing values.

PUBLIC ADMINISTRATIVE VALUES

It has long been noted that a large part of the historical problem of public administration and bureaucracy in the United States results directly from the political community's inability to choose finally among competing values. As Herbert Kaufman conceptualized this situation:

> An examination of the administrative institutions of this country suggest that they have been organized and operated in pursuit successively of three values . . . representativeness, neutral competence, and executive leadership. Each of these values has been dominant (but not to the point of total suppression of the others) in different periods of our history; the shift from one to another generally appears to have occurred as a consequence of the difficulties encountered in the period preceding the change. [Kaufman, 1956: 1057]

Thus, representativeness was stressed during the Jacksonian era (Rosenbloom, 1971, 1977), neutral competence during the period of federal civil service reform (1883) and its aftermath (Rosenbloom, 1971), and executive leadership has been an underlying value in the creation of the modern presidency, which can conveniently be dated from the creation of the Executive Office of the President in 1939 (Kaufman, 1956: 1066). Yet, as Kaufman's analysis suggests, it is not only major systemic changes that are subject to these competing values, the conflict among them is also manifested in personnel developments of more limited scope. For example, throughout the decade of the 1970s there has been intense competition between the values of neutral competence, as manifested in the "merit system," and representation as implemented through "affirmative action" (Rosenbloom, 1977).

James Q. Wilson has also stressed the problem of competing values:

> There is not one bureaucracy problem, there are several, and the solution to each is in some degree incompatible with the solution to every other. First, there is the problem of accountability or

control—getting the bureaucracy to serve agreed-on national goals. Second is the problem of equity—getting bureaucrats to treat cases alike and on the basis of clear rules, known in advance. Third is the problem of efficiency—maximizing output for a given expenditure, or minimizing expenditures for a given output. Fourth is the problem of responsiveness—inducing bureaucrats to meet, with alacrity and compassion, those cases which can never be brought under a single national rule and which, by common human standards of justice or benevolence, seem to require that an exception be made or a rule stretched. Fifth is the problem of fiscal integrity—properly spending and accounting for public money. [Wilson, 1977: 58]

Although Wilson's discussion focuses on the national level, it is also instructive in analyzing the problems confronting urban bureaucracies. However, there are two major differences: (1) competing administrative values on the urban level tend to be associated with ethnic, racial, and religious conflict, and (2) collective bargaining has created an additional power bloc alongside those of "the bureaucracy" and the political-managerial component of urban government—that of public employee unions. Together, these conditions provide urban public personnel administration with high political saliency and complexity.

NEUTRAL COMPETENCE VS. REPRESENTATION: "YANKEES" AND IMMIGRANTS

Historically, at the urban level, the competing values of public administration in the United States have tended to be organized into two kinds of political ethos supported by two diverse, but nonetheless reasonably coherent sets of social groups. As Richard Hofstadter writes:

Out of the clash between the needs of the immigrants and the sentiments of the natives there emerged two thoroughly different systems of political ethics.... One, founded upon the indigenous Yankee-Protestant political traditions, and upon middle class life, assumed and demanded the constant, disinterested activity of the citizen in public affairs, argued that political life ought to be run, to a greater degree than it was, in accordance with general principles and abstract laws apart from and superior to personal needs, and expressed a common feeling that government should be in good part an effort to moralize the lives of individuals while economic life should be intimately related to the stimulation and development of individual character. The other system, founded upon the European background of the immigrants, upon their unfamiliarity with independent political action, their familiarity with hierarchy and authority, and upon the urgent needs that so often grew out of their migration, took for granted that the political life of the individual would arise out of family needs, interpreted political and civil relations chiefly in terms of personal obligations, and placed strong personal loyalties above allegiance to abstract codes of law or morals. [Hofstadter, 1955: 9; as cited in Banfield and Wilson, 1963: 41]

Initially, on the urban scene, this clash was manifested in the struggle for "good government" as opposed to "machine politics." Yankee reformers sought clean, efficient, neutrally competent, and nonpartisan urban governments. Their ethos is represented in such aphorisms as "there is no Democratic or Republican way to pave a street," and "only the best shall serve the state." Whereas machine politicians conceived of urban government mainly as a business; to the Yankees it has been considered primarily a civic duty or responsibility.

The Yankee ethos has had a very substantial impact on urban personnel administration, but it has not altogether eliminated competing values associated with the outlook of other social groups. Yankees were at the forefront of American civil service reform in the late nineteenth century (Rosenbloom, 1971). Their general perspective on this issue was well conveyed by Dorman B. Eaton, one of their leaders:

We have seen a class of politicians become powerful in high places, who have not taken (and who by nature are not qualified to take) any large part in the social and educational life of the people. Politics have tended more and more to become a trade, or separate occupation. High character and capacity have become disassociated from public life in the popular mind. [Eaton, 1880: 392]

The reformers' primary vehicle for overcoming this state of affairs was the introduction of the "merit system." At once this would undermine patronage, the power base of machine politicians, and also encourage efficiency and morality in government.

The appeal of reform has been substantial: By 1900 some 65 cities had adopted merit-oriented personnel administration; by 1930 this number had risen to 250; today, only about 12% of all cities with populations in excess of 50,000 lack comprehensive merit systems; and no U.S. city with a population of 250,000 or more is without some provision for a merit system (Shafritz et al., 1978: 43, 46). However, what the Yankees won in the form of formal regulations has often been circumvented by immigrants and their descendants in the process of implementation.

The Yankee approach to urban politics, with its stress on impersonality and the common good, was antithetical to that of immigrants, whose needs were intensely personal and particularistic. The bulk of immigrants in urban settings, whether Irish, Italian, Jewish, Oriental, or Hispanic, has been subjected to far-reaching discrimination and appalling living conditions. Most observers of the contemporary urban scene are familiar with these problems in relation to today's "minorities," but it is equally crucial to an understanding of urban personnel administration to recognize that similar conditions have long prevailed. The following passage provides an apt description of tenement life in New York City:

> The effect produced upon the mind by an inspection of these human rookeries is a vehement desire to pull down and raze to the ground the vast system which holds in bondage thousands and thousands of men, women, and children. These high brick houses tower up to heaven, each flat holding from five to ten families, and one building numbering frequently a population of six hundred souls.... [T]he condition of the Irish is by no means the worst, but the atmosphere of the place is death, morally and physically. Crowded into one small room a whole family lives, a unit among a dozen other such families.... The street below is dirty and ill kept. In the basement is a beer saloon, where crime and want jostle each other, and curses fill the air. On the other side is an Italian tenement reeking with dirt and rags. Close by is a Chinese quarter, or a Polish-Jew colony. Everywhere the moral atmosphere is one of degradation and human demoralization. Gross sensuality prevails. The sense of shame, if ever known, is early stifled. Domestic morals are too often abandoned and simple manners are things of the past. There is no family life possible in such surroundings.... The fireside is hired by the week, the inmate is a hireling, and his family are most probably chained as hirelings also in some great neighbouring factory or mill. (LaGumina and Cavaioli, 1974: 93; quoting Abbott, 1926]

Unequal pay for equal work was prevalent, intermarriage between immigrants and Yankees was rare, intergroup violence was common, and even lynchings of immigrants were not unheard of (LaGumina and Cavaioli, 1974). Nor was this situation confined to New York, despite its historical status as the major port of entry to American life. Although the great waves of immigration that brought some 23.5 million people to the United States from 1881 to 1920 had declined substantially in the following decades (LaGumina and Cavaioli, 1974: appendix, 354-355), as late as 1960, 30% or more of the populations of the following cities with at least 500,000 inhabitants were either foreign born or had at least one foreign-born parent: New York (48.6%), Boston (45.5%), San Francisco (43.5%), Chicago (35.9%), Buffalo (35.4%), Los Angeles (32.6%), Detroit (32.2%), Seattle (31.4%), Cleveland (30.9%), Pittsburgh (30.3%), and Milwaukee (30.0%) (Banfield and Wilson, 1963: 39, table 3). If "undocumented" aliens were added to the totals, additional Southwestern cities, including San Antonio and San Diego, could undoubtedly be added to the list.

The immigrant ethos, with its stress on immediate tangible rewards, jobs, and group recognition as opposed to some abstract notion of the common civic good, fit hand in glove with the rise of the political machine. It is most useful to understand the latter as a business organization of a particular variety. As Banfield and Wilson write:

> The machine . . . is apolitical: it is interested only in making and distributing income—mainly money—to those who run it and work for it. Political principle is foreign to it, and represents a danger and a threat to it. . . . The existence of the machine depends upon its ability to control votes. This control becomes possible when people place little or no value on their votes, or, more precisely, when they place a lower value on their votes than they do on the things which the machine can offer them in exchange for them. The voter who is indifferent to issues, principles, or candidates puts little or no value on his vote and can be induced relatively easily (or cheaply) to put it at the machine's disposal. [Banfield and Wilson, 1963: 116-117]

And, perhaps needless to add, it was from immigrants and other disadvantaged populations, rather than from middle-class Yankees, that such votes were most likely.

Thus, the immigrant had two commodities to sell: labor and votes. The latter often represented the more viable strategy for upward mobility and assimilation under prevailing conditions. In his classic essay on "The Latent Functions of the Machine," Robert Merton wrote, a "set of distinctive functions fulfilled by the political machine for a special subgroup is that of providing alternative channels of social mobility for those otherwise excluded from the more conventional avenues for personal 'advancement'" (Merton, 1961: 185). Frances Fox Piven elaborates upon this function:

> Municipal jobs have always been an important resource in the cultivation of political power. As successive waves of immigrants settled in the cities, their votes were exchanged for jobs and other favors, permitting established party leaders to develop and maintain control despite the disruptive potential of new and unaffiliated populations. The exchange also facilitated the integration of immigrant groups into the economic and political structures of the city, yielding them both a measure of influence and some occupational rewards. Public employment was a major channel of mobility for the Italian, the Irish and the Jew, each of whom, by successively taking over whole sectors of the public services, gave various municipal agencies their distinctly ethnic coloration. [Piven, 1973: 380]

Thus, "patronage is peculiarly important for minority groups, involving much more than the mere spoils of office. Each first appointment given a member of any underdog element is a boost in that element's struggle for social acceptance" (Lubell, 1955: 81).

Under these historical conditions, it is not surprising that the immigrant ethos has opposed any public personnel process that might be used to deny jobs to members of their own particular groups. Consequently, when the Yankee reformers were able to introduce civil service reform, the immigrant ethos denounced it as "the biggest fraud of the age.... [T]he curse of the nation" (Shafritz et al., 1978: 44; quoting *Plunkitt of Tammany Hall*). Moreover, in many places, such as Chicago, political bosses were able to circumvent formal merit systems to such an extent that civil service regulations were rendered almost meaningless (Shafritz et al., 1978: 51-59).

Similarly, however, after ethnics solidified their hold on urban public employment, whether through patronage or merit, and as they became increasingly assimilated and absorbed by the larger society, they were sometimes found in the forefront of opposition to any effort to dilute the merit system, especially through "affirmative action." To a considerable extent, this opposition has been based on an understanding of the importance of public employment in the upward mobility of ethnic groups. Any attempt at increasing the representation of today's minorities may be viewed as a threat to a social-political position attained by white ethnic groups only after protracted struggle. This is especially true under conditions of retrenchment or nongrowth which may require general sacrifices or that gains for one group be balanced out by losses for another. Thus, just as the civil service reformers of the late nineteenth century viewed the introduction of the merit system as a means of bringing about widespread political change, today many of its defenders among white ethnics view that system as a means of maintaining their own group's power and position (Bent and Reeves, 1978: 33-34). The New York City teachers' strike of 1968, which led to the polarization of the city, its employees, and their unions along racial and ethnic lines represents an excellent, though extreme, case in point (Raskin, 1972: 134-135).

However, it would be too facile to dismiss the controversy over affirmative action as solely a battle over jobs. It is also a contest over principle. As they moved from "immigrants" to "ethnics," these groups became socialized to prevailing American perspectives and their outlook on political life underwent substantial changes:

> Possibly because it lacked the excitement of the Indian wars or because it still is so close to us, the saga of this twentieth-century odyssey of America's urban masses has gone unsung. . . .
>
> For the urban masses each advance into a new neighborhood has also been a "beginning over again," which took them ever further from their European origins in the case of the immigrants, or, with Negroes, from the Plantation South. There has been much pooh-poohing of social climbing, without appreciation of the fact that it is a vital part of the Americanization process. The move to a "nicer" neighborhood would often be celebrated by . . . [Americanization].
>
> Because they spent their youth in rootless tenements which knew no community life, they have been buying homes and have become *doubly civic-minded* in their eagerness to build a community in which their children might escape the deprivations of their own childhood. [Lubell, 1955: 65-67; emphasis added]

Undoubtedly, therefore, much of the current support for reformist-oriented public personnel administration coming from white ethnics is also based on the community-oriented belief that such practices are desirable for the society generally. Remarkably, the United States has moved so far away from the kind of personnel administration that was widely prevalent in the heyday of the political machine, that in 1976 the Supreme Court actually held patronage dismissals from run-of-the-mill public service jobs to be an unconstitutional infringement upon rights protected under the First and Fourteenth Amendments (Elrod vs. Burns, 1976).

In sum, out of the struggle between Yankee and immigrant emerged the fundamental personnel values of neutral competence and representation. Although both of these values were partly rationalizations of efforts aimed at creating a particular distribution of political power, they nevertheless have become widely accepted and manifested in public personnel policy generally. After several generations, what were once competing values were synthesized through the experience of white ethnics in urban America. Merit and representation coexisted in relative harmony until additional groups voiced old demands in a new fashion.

NEUTRAL COMPETENCE VS. REPRESENTATION: "AFFIRMATIVE ACTION"

By 1970 the values of neutral competence, as manifested in the merit system, and representation were well entrenched in American public personnel administration. Indeed, public personnel policy in the preceding three decades assumed that merit and representation were fully compatible (Rosenbloom, 1977). Thus, until the 1970s, equal employment opportunity policy was premised on the belief that if artificial barriers to the employment of blacks and Hispanics were removed, members of these groups would readily obtain a substantial degree of representation through the merit system. Consequently, non-discrimination and complaint systems were the focus of EEO programs. However, as the black power movement of the 1960s shifted minority objectives from equal opportunity to equality per se (Christenson, 1970: Ch. 2), the merit system came under increasing challenge as neither neutral nor a viable means of determining competence.

By 1970 several large and medium-sized U.S. cities had acquired substantial minority populations. For example, blacks comprised 71% of the population of Washington, DC, 51% of Atlanta, 54% of Newark, 53% of Gary, 46% of Baltimore, 44% of Detroit, 41% of St. Louis, 33% of Chicago, 21% of New York, and 34% of Philadelphia (Department of Commerce, 1977: 22-24). Where black machine politicians were effective, the black route to public employment was not much different from that of the immigrants of generations past. As Banfield and Wilson observe, "In a city with a partisan, ward-based machine, Negroes will be organized as a sub-machine" maintained "in the usual way, by exchanging jobs, favors, and protection for votes" (Banfield and Wilson, 1963: 304). However, by the time that blacks had become a substantial political force on the urban scene, the amount of patronage jobs extant had drastically declined:

> Now blacks are the newcomers. But they come at a time when public employment has been preempted by older groups and is held fast through civil-service provisions and collective bargaining contracts. Most public jobs are no longer allocated in exchange for political allegiance, but through a "merit" system based on formal qualifications. [Piven, 1973: 380]

This would not be problematic from the perspectives of the value of representation were it not for the fact that the merit system does not produce a high degree of minority employment in middle- and especially upper-level public sector jobs.

Blacks and other minority group members often find cities to be the most receptive public employers (Kranz, 1976: 177). Nevertheless, the following statements serve to describe their urban public employment patterns:

- Cities as a group are below the national parity ratios for Spanish, Asian, Indian, and female minorities.
- Blacks make up a large percentage of public welfare and hospital employees, but only 3.3% of those in fire protection.
- Minorities employed by the nation's cities are inordinately compressed into a few occupational categories and grossly under-represented in the official/administrator class.
- With the exception of Asians, minorities are disproportionately crammed in the lower-pay brackets, whereas whites and males occupy the better-paying, higher-level positions.

> A third of those earning less than $4,000 are minority....
> [H]owever, minorities are only 8.5% of those earning more
> than $16,000. [Kranz, 1976: 177, 180, 180, 182, 184]

Based upon these data, it would not be difficult to build a prima facie case that urban public personnel administration is discriminatory in its harsh impact upon minorities.

It is not surprising, therefore, that minorities in a number of jurisdictions have challenged prevailing public personnel practices in court on the grounds that they violate the equal protection clause of the Fourteenth Amendment and/or the provisions of the Equal Employment Opportunity Act of 1972. In 1976 the Supreme Court outlined how cases arising under the EEO Act are to be treated:

> Under Title VII, Congress provided that when hiring and promotion practices disqualifying substantially disproportionate numbers of blacks are challenged, discriminatory purpose need not be proved, and that it is an insufficient response to demonstrate some rational basis for the challenged practices. It is necessary, in addition, that they be "validated" in terms of job performance.... However this process proceeds, it involves a ... probing judicial review of, and less deference to, the seemingly reasonable acts of administrators and executives. [Washington vs. Davis, 1976: 4794]

The requirement of validation has proven difficult for contemporary merit-oriented public personnel administration. Put simply, for a variety of technical reasons it is (1) difficult to show a substantial level of validity even when an examination possesses this quality, and (2) such validity is decidedly not always or perhaps even generally present (Rosenbloom and Obuchowski, 1977). Indeed, validity studies generally arrive at coefficients that would explain no more than about 10% of the employee's on-the-job performance (McClung, 1968: 340, table 2). On the other hand, a substantial number of merit exams manifest a harsh racial bias (Rosenbloom and Obuchowski, 1977; Rosenbloom, 1977).

Thus, the "merit system" is by no means always neutral. Nor can it be safely assumed that it yields a high level of competence. As one court expressed it, "by definition," unvalidated selection procedures, "have not been shown to be predictive of successful job performance. Hence there is no reliable way to know that any accepted applicant is truly better qualified than others who have been rejected" (NAACP vs. Allen, 1974: 618). Under these circumstances, the case for representation through affirmative action is enhanced. "Until ... selection procedures ... have been properly validated, it is illogical to argue that quota hiring produces unconstitutional 'reverse' discrimination, or a lowering of employment standards, or the appointment of less or unqualified persons" (NAACP vs. Allen, 1974: 618).

Hence, in addition to philosophic arguments in favor of affirmative action there has been a legal logic to it. In fact, several courts have imposed minority hiring quotas on local jurisdictions as a remedy for past discriminatory practices resulting from the harsh racial impact of merit examinations (Rosenbloom and Obuchowski, 1977). Although a substantial majority of the population has registered its distaste for quota approaches (Bent and Reeves, 1978: 32), thus far, they can be interpreted as both legal and constitutional *as a remedy for past discrimination.* While invalidating a quota system for admissions to medical school in Board of Regents vs. Bakke (1978), the Supreme Court did not directly confront the question of the appropriateness of quotas as a remedy, although five justices seemed willing to support their use for this purpose. In any event, the minority assertion of the value of representation in challenge to that of neutral competence has severely damaged the legitimacy of the merit system in the absence of adequate validation. Regardless of whether the imposition of quotas continues, invalid and unvalidated examinations having a harsh racial impact are certain to be cast aside in favor of other, less biased personnel procedures.

But of what will these procedures consist? This will be a major question in the 1980s. Several possibilities are available, although, based on current Supreme Court rulings, the widespread use of quotas (except as a remedy) and patronage are exceedingly unlikely. Among other techniques, the standard use of the "rule of three" might be relaxed to enable appointing officials to have a wider choice from among qualified applicants. Significantly, this approach has been endorsed by the National Civil Service League and is already used in some jurisdictions (Rosenbloom, 1977: 141-142). Another approach would be to use a pass/fail cut-off on examinations and select randomly form among those deemed basically qualified for the available positions (Rosenbloom, 1972). Either approach would mitigate the harsh racial impact of contemporary merit examinations. Finally, it may yet prove possible to develop valid and nondiscriminatory examinations, although this has so far proved to be an elusive goal. Clearly, the agenda for public personnel administration in the 1980s would be substantial even were it not for the impending likelihood of substantial retrenchment. The latter, however, will serve to intensify conflict.

PUBLIC PERSONNEL ADMINISTRATION UNDER RETRENCHMENT

The problem of competing values in public administration has long been evident in the United States. In terms of urban public personnel administration the most salient conflict has involved the values of representation and neutral competence. This struggle has been associated with the polarization of ethnic and racial groups and has been of such an intensity as to sometimes engender violence. This intensity will be further enhanced under conditions of declining job opportunities in the public sector. Consequently, a major set of issues for public personnel administration in the 1980s will go beyond the questions of how representation and neutral competence can be joined in hirings and promotions; the question of how these values should be manifested in cutbacks will also come to the fore.

Public personnel administration in the United States has comparatively little experience with large-scale reductions in force, despite some recent and well publicized cutbacks such as the 21% decline in New York City's workforce from 1975 to 1977 (Glasberg, 1978: 327). As Charles Levine observes:

> We know very little about the decline of public organizations and the management of cutbacks. This may be because even though some federal agencies . . . and many state and local agencies have expanded and then contracted, or even died, the public sector as a whole has expanded enormously over the last four decades. In this period of expansion and optimism among proponents of an active government, isolated incidents of zero growth and decline have been considered anomalous; and the difficulties faced by the management of declining agencies coping with retrenchment have been regarded as outside the mainstream of public management concerns. [Levine, 1978: 316]

It is widely believed, however, that in general,

> lack of growth . . . creates a number of serious personnel problems. For example, the need to reward managers for directing

organizational contraction and termination is a problem because without growth there are few promotions and rewards available to motivate and retain successful and loyal managers—particularly when compared to job opportunities for talented managers outside the declining organization. Also, without expansion, public organizations that are constrained by merit and career tenure systems are unable to attract and accommodate new young talent. Without an inflow of younger employees, the average age of employees is forced up, and the organization's skill pool becomes frozen at the very time younger, more flexible, more mobile, less expensive and (some would argue) more creative employees are needed. [Levine, 1978: 317]

Unfortunately, though, contemporary cutback strategies often exacerbate these problems and those of attaining a representative public sector.

Among the most common strategies for dealing with personnel retrenchment are:

(1) *Using seniority as a decision rule.* Firing the least senior first may have little to do with equity, efficiency, or effectiveness. However, it is directed at "the need to provide secure career-long employment to neutrally competent civil servants" (Levine, 1978: 321). Since more senior personnel are often older as well, retention based upon seniority may reduce the overall problem of future employability for dismissed personnel. In many urban settings, the use of seniority as a decision rule would have a substantial and adverse impact upon the employment of women and minorities, especially in the middle and upper levels, because these groups have often been among those hired most recently, often as a result of contemporary equal employment opportunity efforts. Here, again there is a conflict between the values of neutral competence and representation. Despite their harsh racial and gender impacts, bona fide seniority systems are likely to be upheld in court unless they perpetuate illegal discrimination, that is, discrimination occurring after and in contravention to the 1972 EEO Act (Cebulski, 1977: 21-24).

(2) *Allowing reduction through attrition.* This is perhaps the least dramatic, painful, and conflictual way of reducing the size of the public sector workforce. However, since the mandatory retirement age has recently been raised to seventy throughout the society generally (and unlimited at the federal level), attrition may also prove to be among the slowest techniques available. Moreover, because the state of human

resources planning is so primitive (Shafritz et al., 1978: 71-92), attrition is not subject to sufficiently adequate prediction or organizational planning. Consequently, it may result in a surplus of some skills and a deficit of others.

(3) *Veteran preference.* The federal government and many other public jurisdictions provide a set of preferences to veterans which may include special retention rights during reductions in force. Such provisions are not based upon concepts of efficiency, effectiveness, or other widely held personnel values. They simply result from a desire to use public employment for extraorganizational social and political reasons. Their impact, however, may be substantial in terms of the representation of women and members of minority groups because "American veterans are 98 per cent male and 92 per cent nonminority" (Campbell, 1978: 102).

(4) *Productivity.* In the abstract, basing reductions in force upon productivity measures would appear to make a great deal of sense. The least productive employees could be dismissed before the more productive ones. Of course, as suggested in connection with seniority above, governments, unlike private employers, must be acutely concerned with what happens to personnel removed from the public service. The least productive may also be the least reemployable, in which case the solution to one problem would simply worsen another. More substantially, however,

> productivity analysis in the public sector has not proved to be easy, and the reason for this lies in the intractable nature of the measurement difficulties that are encountered. Public output, characterized to a greater or lesser degree by nonexlusion and nonrivalry, cannot be quantified in discrete units. In consequence, those who have worked in this vineyard have very often emerged with a mishmash of "productivity indicators" that cannot be compared, aggregated, or utilized effectively for expenditure control. [Bahl and Burkhead, 1977: 267]

Hence, attempting to use productivity as a decision rule in reductions in force is likely to prove inoperable or lead to perverted results.

Where does this leave the state of the art? Since none of the commonly used approaches for making reductions in force is wholly satisfactory or unequivocally better than any other, and since, insofar as they are operational, they tend to have differential impacts on various social groups, it is likely that this will be a major area of political contention during the 1980s and that politics will have a great deal to do with the implementation of personnel cutbacks. Veterans will want to retain their preferences; more senior members will stress the desirability of seniority systems; current employees may stress the potentials of attrition; minority group members, women, and other groups such as the handicapped are likely to press for some sort of social preference; outsiders will continue to clamor for productivity as a decision rule. As in the past, the struggle over city personnel will take place in several arenas, including the realms of elective officials and professional public personnel administrators. What has changed, however, is that now another and often very powerful participant is on the scene—the public employee labor union.

LABOR RELATIONS AND RETRENCHMENT

It is evident that public employee labor unions and associations will have a large role in the public personnel administration of the 1980s. Here again, competing values are present. On the one hand, there is a desire for establishing and maintaining accountability for elective officials. This objective generally entails the supporting proposition that politically appointed managers should be firmly in control of public employees. The Federal Civil Service Reform Act of 1978 presents a contemporary example of this approach. On the other hand, however, there is the competing value of worker participation. While less entrenched than the value of accountability, participation has emerged as an integral part of public personnel administration (Berkley, 1971; 1975). For instance, Executive Order 11491 (1969), dealing with federal labor relations, includes the following passage:

> The well-being of employees and efficient administration of the Government are benefited by providing employees an opportunity to participate in the formulation and implementation of personnel policies and practices affecting the conditions of their employment.

Although there has been much opposition to participatory public personnel administration on the part of local and other public officials, there is little doubt that labor unions have made major inroads on what were once considered management prerogatives (Nigro and Nigro, 1977).

Indeed, worker participation has advanced partly out of recognition that as a result of their professional expertise, public employees often have a major contribution to make to the formulation and implementation of public policy. As Nigro and Nigro point out:

> The evidence is clear that, in government, professionalism and unionism, far from being antithetical, are mutually reinforcing: the union helps to achieve professional goals. Many professionals choose government employment in order to be able to contribute public service; loyalty to profession is combined with zeal to help the community. They identify with the mission of their organizations with an intensity seldom found in professionals working for private entities. And this is why the management rights concept is threatened by collective bargaining much more in government than in the private sector. [Nigro and Nigro, 1977: 141]

Moreover, in an age of low voter turnouts and ambiguous "mandates" from the electorate, the traditional concept of accountability is placed under considerable strain: "The ultimate management is the electorate, and employees at all levels of the organization have a responsibility for serving the public" (Nigro and Nigro, 1977: 141).

By 1976 more than half of all local government employees belonged to a labor union (Shafritz et al., 1978: 198) and it was on this level that labor relations had become most problematic. Discord and work stoppages were common (Lewin, 1977: 149). As a result of recession, past patterns of labor relations, and other factors, the fiscal condition of urban America became increasingly perilous. Retrenchment, once looked upon as a temporary condition at worst, now began to appear as a permanent aspect of urban life.

Thus far, the reaction of municipal unions to retrenchment has not been encouraging from the perspectives of harmonious public personnel administration. As David T. Stanley observes:

> Fiscal crises strengthened the hand of management as tough bargainer, job abolisher, and antifeatherbedder. The unions tried hard to respond. They made strong public statements urging pay increases and opposing job reductions. They struck for more money (unsuccessfully in San Francisco and with modest success in Cleveland), struck to protest an increase in duty requirements (New York City schoolteachers), and struck to protest layoffs

(New York sanitation men). They objected to city work being done by CETA employees (Buffalo and New York). [Stanley, 1977: 32]

Retrenchment has further complicated municipal labor relations by weakening their leadership and blurring union goals and strategies (Horton, 1976: 198). Yet, at least one issue was clarified: "No longer were resources viewed as being capable of being easily distributed to one group without hurting the perceived interests of another. By 1975 [New York City] civil servants realized that wage increases meant increased layoffs, and the public understood that wage increases meant reduced services" (Horton, 1976: 201).

The redistributive nature of urban retrenchment does not bode well for representative public services. Seniority is emerging as the unions' preferred strategy for determining who will be retained when reductions in force are necessary:

> In some public agencies, the lack of seniority protections or of any rules governing dismissals for economic reasons, have brought demands for negotiated protections. When financial stringencies begin to weaken the traditionally assumed security of public service jobs, the negotiating table becomes the arena for seeking a direct remedy for threatened security. Employees still turn to legislative or administrative lobbying, but often more to prevent erosions of existing protections, prompted by employers' attempts to modify seniority provisions. [Cebulski, 1977: 11]

But, as mentioned earlier, the seniority criterion will often have a harsh impact on the employment of women and members of minority groups. This, in turn, may lead to even further polarization over personnel issues.

A tension between municipal labor unions and minorities has long been evident:

> The black community often has a dual complaint against the unions. On the one hand unions have been a party to exclusion of minorities for police, fire, and teaching jobs; and on the other hand, white union members have been heavy-handed and insensitive in delivering municipal services to low-income blacks.... [T]he fact that some unions in the public sector with an increasingly large black membership retain lily-white leadership at the national levels is irritating. [Gould, 1972: 151]

Indeed, in some occupations, such as police, minorities have sometimes formed their own employee associations. Under redistributive conditions, this tension may develop into a full-fledged conflict over: (1) the leadership of unions, (2) the extent to which minorities will be able to form separate unions or associations, (3) the techniques of retrenchment, (4) affirmative action, and (5) the use of residency requirements to insure that municipal employees live within the boundaries of their employer's jurisdiction. The constitutionality of the latter has been upheld (McCarthy vs. Philadelphia Civil Service Commission, 1976), and their imposition will often be favored by members of minority groups.

PUBLIC PERSONNEL ADMINISTRATION IN THE 1980s

Toward the end of the 1960s, public personnel administration was held in exceptionally low repute. In a rank order list of 27 fields of specialization among American political scientists, it emerged in last place. Some referred to it as a "dying, if not already dead, specialization" (Shafritz, 1973: 6-7). Among practitioners of public administration, personnelists were also held in low regard (Rosenbloom, 1977: 81-86, 100-101, note 16). As the 1970s progressed, however, increasing attention was paid to public personnel administration. Labor relations, equal employment opportunity, the constitutional position of public employees, civil service reform, and issues such as "whistleblowing" could be found among the important issues on the nation's political agenda. It is likely that public personnel administration will remain of high salience during the 1980s, especially in the urban sector. As we proceed through the coming decade, therefore, it is critical that in rethinking personnel issues, personnelists and urbanologists be continually cognizant of "the 'multiplier' importance of public service—great changes in a wide arena are instigated by small alterations in governmental personnel policy" (Krislov, 1967: 5).

REFERENCES

ABBOTT, E. (1926) Historical Aspects of the Immigration Problem. Chicago: University of Chicago Press.

BAHL, R. and J. BURKHEAD (1977) "Producitvity and the measurement of public output," pp. 253-270 in C. Levine (ed.) Managing Human Resources. Beverly Hills: Sage Publications.

BANFIELD, E. and J. WILSON (1963) City Politics. Cambridge, MA: Harvard University Press and MIT Press.

BENT, E. and T. REEVES (1978) Collective Bargaining in the Public Sector. Menlo Park, CA: Benjamin/Cummings.

BERKLEY, G. (1975) The Craft of Public Administration. Boston: Allyn and Bacon. Board of Regents vs. Bakke (1978). 46 Law Week 4896.

——— (1971) The Administrative Revolution. Englewood Cliffs, NJ: P entice-Hall.

CAMPBELL, A. (1978) "Civil service reform: a new commitment." Public Admin. Rev. 38 (March/April): 99-103.

CEBULSKI, B. (1977) Affirmative Action Versus Seniority–Is Conflict Inevitable? Berkeley: Institute of Industrial Relations, University of California.

EATON, D. (1880) The Civil Service in Great Britain. New York: Harper and Row.

Elrod vs. Burns (1976) 427 U.S. 347.

GLASBERG, A. (1978) "Organizational responses to municipal budged decreases." Public Admin. Rev. 38 (July/August): 325-331.

GOULD, W. (1972) "Labor relations and race relations," pp. 147-159 in S. Zagoria (ed.) Public Workers and Public Unions. Englewood Cliffs, NJ: Prentice-Hall.

HOFSTADTER, R. (1955) The Age of Reform. New York: Alfred A. Knopf.

HORTON, R. (1976) "Economics, politics, and collective bargaining: the case of New York City," pp. 183-202 in Institute for Contemporary Studies (ed.) Public Employee Unions. San Francisco: Institute for Contemporary Studies.

KAUFMAN, H. (1956) "Emerging Conflicts in the Doctrines of Public Administration." Amer. Pol. Sci. Rev. 50 (December): 1057-1073.

KRANZ, H. (1976) The Participatory Bureaucracy. Lexington, MA: Lexington Books.

KRISLOV, S. (1967) The Negro in Federal Employment. Minneapolis: University of Minnesota Press.

LaGUMINA, S. and F. CAVAIOLI (1974) The Ethnic Dimensions in American Society. Boston: Holbrook.

LEVINE, C. (1978) "Organizational decline and cutback management." Public Admin. Rev. 38 (July/August): 316-324.

LEWIN, D.s(1977) "Collective bargaining and the right to strike," pp. 145-164 in Institute for Contemporary Studies (ed.) Public Employee Unions. San Francisco: Institute for Contemporary Studies.

LUBELL, S. (1955) The Future of American Politics. Garden City, NY: Doubleday Anchor.

McCarthy vs. Philadelphia Civil Service Commission (1976). 47 L. Ed. 2d 366.

McCLUNG, G. (1968) "Statistical techniques in testing," pp. 335-354 in J. Donovan (ed.) Recruitment and Selection in the Public Service. Chicago: Public Personnel Association.

MERTON, R. (1961) "The Latent Functions of the Machine," pp. 180-190 in E. Banfield (ed.) Urban Government. New York: Free Press.

NAACP vs. Allen (1974) 493 F2d. 614.
NIGRO, F. and L. NIGRO (1977) "Public sector unionism," pp. 141-158 in C. Levine (ed.) Managing Human Resources. Beverly Hills: Sage Publications.
PIVEN, F. (1973) "Militant civil servants in New York City," pp. 380-385 in W. Burnham (ed.) Politics/America. New York: Van Nostrand Reinhold.
RASKIN, A. (1972) "Politics up-ends the bargaining table," pp. 122-146 in S. Zagoria (ed.) Public Workers and Public Unions. Englewood Cliffs, NJ: Prentice-Hall.
RIORDON, W. (1948) Plunkitt of Tammany Hall. New York: Alfred A. Knopf.
ROSENBLOOM, D. H. (1977) Federal Equal Employment Opportunity. New York: Frederick A. Praeger.
——— (1972) "Federal equal employment opportunity: another strategy." Personnel Admin. and Public Personnel Rev. 1 (July/August): 38-41.
——— (1971) Federal Service and the Constitution. Ithaca, NY: Cornell University Press.
——— and C. OBUCHOWSKI (1977) "Public personnel examinations and the Constitution: emergent trends." Public Admin. Rev. 37 (January/February): 9-18.
SHAFRITZ, J. (1973) Position Classification. New York: Frederick A. Praeger.
———, W. BALK, A. HYDE, and D. H. ROSENBLOOM (1978) Personnel Management in Government. New York: Marcel Dekker.
STANLEY, D. (1977) "The ambiguous role of the urban public employee," pp. 23-35 in C. Levine (ed.) Managing Human Resources. Beverly Hills: Sage Publications.
THOMPSON, F. (1978) Personnel Policy in the City. Berkeley: University of California Press.
——— (1977) "Institutional barriers to equity in local government," pp. 83-111 in C. Levine (ed.) Managing Human Resources. Beverly Hills: Sage Publications.
U.S. Department of Commerce (1977) Statistical Abstract. Washington, DC: Bureau of the Census.
——— (1976) Bicentennial Statistics. Washington, DC: Bureau of the Census.
WILSON, J. (1977) "The bureaucracy problem," pp. 57-62 in A. Altshuler and N. Thomas (eds.) The Politics of the Federal Bureaucracy. New York: Harper and Row.

THE AUTHORS

RICHARD D. BINGHAM is Associate Professor of Political Science and Chairperson of the Department of Urban Affairs at the University of Wisconsin-Milwaukee. Dr. Bingham's books include *Public Housing and Urban Renewal: Analysis of Federal Local Relations* and *The Adoption of Innovation by Local Government.* His articles have appeared in such journals as *Urban Affairs Quarterly* and *Policy Studies Journal.*

JOHN P. BLAIR is an economist in the Department of Housing and Urban Development, Office of Community Planning and Development on leave from the Department of Urban Affairs and the School of Business Administration, University of Wisconsin-Milwaukee. His articles on Urban Development and Finance have appeared in *Public Choice, Growth and Change* and *Land Economics* among other places. He recently completed a textbook titled *Real Estate Analysis,* coauthored with G. V. Barrett.

WARNER BLOOMBERG, Jr. is an Associate Professor of Sociology, San Diego State University. His research interests include urban social systems and minority groups. Professor Bloomberg was coeditor of *Power, Poverty, and Urban Policy* and *The Quality of Urban Life.*

TERRY NICHOLS CLARK is Associate Professor of Sociology at the University of Chicago, and has taught at Columbia, Harvard, Yale, and the Sorbonne. He has been a consultant to the Urban Institute and the Department of Housing and Urban Development. He is the author or coauthor of *Community Structure and Decision Making* (1968), *Community Politics* (1971), *Community Power and Policy Outputs* (1973), and *Comparative Community Politics* (1974).

GARY GAPPERT is an economist and Director of Urban Development, Research for Better Schools. His interests are Urban Development and Manpower Policy. Dr. Gappert is author of *Post Affluent America* and coeditor of *The Social Economy of Cities*.

ANN LENNARSON GREER is Director of the Urban Research Center and Associate Professor of Urban Affairs and Sociology at the University of Wisconsin-Milwaukee. Previously she explored issues of health policy with the Health Resources Administration. Her most recent book is *The Mayor's Mandate: Municipal Statecraft and Political Trust*.

SCOTT GREER is Professor of Sociology and Urban Affairs at the University of Wisconsin-Milwaukee. His interests include the study of complex organizations and urban social institutions. His published works include *Governing the Metropolis*, *The Logic of Social Inquiry*, and *The Urban View*.

F. TED HEBERT is a Visiting Professor at the University of North Carolina, Chapel Hill. He has published widely in journals such as *Urban Affairs Quarterly*, *Public Administration Review*, and *The American Journal of Political Science*. He is coauthor of *The Politics of Raising State and Local Revenues*.

WAYNE LEE HOFFMAN is a political scientist and Research Associate with The Income Security and Pension Policy Group at The Urban Institute. In Addition to his interest in the organization of citizens' policy views, his past research has included evaluation of welfare and tax policy reforms. He is currently developing simulation modeling techniques for assessing the distributional impact of the changing cost and availability of principal energy forms.

THEODORE J. LOWI is the John L. Senior Professor of American Institutions at Cornell University. He received his B.A. from Michigan

State University and his Ph.D. from Yale. Among his books are *Private Life and Public Order* and the classic *The End of Liberalism*.

HENRY MAIER is Mayor of Milwaukee, Wisconsin. He is the only mayor to have been President of the U.S. Conference of Mayors, the National League of Cities and the U.S. Conference of Democratic Mayors. Mayor Maier is author of *The Challenge to the Cities*.

LOUIS H. MASOTTI is Professor of Political Science and Urban Affairs and Director, Center for Urban Affairs at Northwestern University. His interests include urban politics, suburbs and suburbanization, and comparative systems. Recent books by Dr. Masotti are *The New Urban Politics* and *The City in Comparative Perspective*.

JOHN L. MIKESELL is Professor and Chairman of the Faculty of Economics and Finance in the School of Public and Environmental Affairs, Indiana University. He holds a Ph.D. in economics from the University of Illinois and is actively involved in research in state and local government finance, public choice, and the economics of regulation. His research appears in such journals as *Public Finance Quarterly, Public Finance/Finances Publiques, National Tax Journal, Public Choice, Public Administration Review,* and *The Quarterly Review of Economics and Business*.

CHAVA NACHMIAS is an Assistant Professor of Urban Affairs and Sociology at the University of Wisconsin-Milwaukee. She has published in the areas of urban education, ethnic relations, and methodology. Her recent articles appeared in *Urban Education, American Journal of Sociology,* and *International Journal of Comparative Sociology*.

DAVID NACHMIAS is Professor of Urban Affairs and Political Science at the University of Wisconsin-Milwaukee. He has published in the fields of public policy and methodology. His recent book is *Public Policy Evaluation* (1978).

HAROLD M. ROSE is Professor of Urban Affairs and Geography at the University of Wisconsin-Milwaukee. He was awarded the Van Cleef Memorial Medal by the American Geographical Society for his contributions to urban geography and is a past president of the Association of

American Geographers. He has published numerous books and articles and is currently researching lethal aspects of urban violence.

DAVID H. ROSENBLOOM is Professor of Public Administration at The Maxwell School, Syracuse University. A previous contributor to Urban Affairs Annual Reviews, he has written widely in the areas of public personnel and bureaucratic politics. His latest book, coauthored with David Nachmias, is *Bureaucratic Culture: Citizens and Administrators in Israel.*